Birds of
The Indian Hills

Birds of
The Indian Hills

Douglas Dewar

Vij Books India Pvt Ltd
New Delhi (India)

Published by

Vij Books India Pvt Ltd
(Publishers, Distributors & Importers)
2/19, Ansari Road
Delhi – 110 002
Phones: 91-11-43596460, Mob: 98110 94883
e-mail: contact@vijpublishing.com
web: www.vijbooks.in

Copyright © 2022

ISBN: 9789395675291

Contents

PART I

Birds of the Himalayas

INTRODUCTION

The avifauna of the Himalayas is a large one. It includes birds found throughout the range, birds confined to the eastern or western portions, birds resident all through the year, birds that are mere seasonal visitors, birds found only at high elevations, birds confined to the lower hills, birds abundant everywhere, birds nowhere common. Most ornithological books treat of all these sorts and conditions of birds impartially, with the result that the non-ornithological reader who dips into them finds himself completely out of his depth.

He who plunges into the essays that follow need have no fear of getting out of his depth. With the object of guarding against this catastrophe, I have described as few birds as possible. I have ignored all those that are not likely to be seen daily in summer in the Himalayas at elevations between 5000 and 7000 feet above the sea-level. Moreover, the birds of the Western have been separated from those of the Eastern Himalayas. The result is that he who peruses this book will be confronted with comparatively few birds, and should experience little difficulty in recognising them when he meets them in the flesh. I am fully alive to the fact that the method I have adopted has drawbacks. Some readers are likely to come across birds at the various hill stations which do not find place in this book. Such will doubtless charge me with sins of omission. I meet these charges in anticipation by adopting the defence of the Irishman, charged with the theft of a chicken, whose crime had been witnessed by several persons: "For every witness who saw me steal the chicken, I'll bring twenty who didn't see me steal it!"

The reader will come across twenty birds which the essays that follow will enable him to identify for every one he sees not described in them.

THE HABITAT OF HIMALAYAN BIRDS

Himalayan birds inhabit what is perhaps the most wonderful tract of country in the world. The Himalayas are not so much a chain of mountains as a mountainous country, some eighty miles broad and several hundred long—a country composed entirely of mountains and valleys with no large plains or broad plateaux.

There is a saying of an ancient Sanskrit poet which, being translated into English, runs: "In a hundred ages of the gods I could not tell you of the glories of Himachal." This every writer on things Himalayan contrives to drag into his composition. Some begin with the quotation, while others reserve it for the last, and make it do duty for the epigram which stylists assure us should terminate every essay.

Some there are who quote the Indian sage only to mock him. Such assert that the beauties of the Himalayas have been greatly exaggerated—that, as regards grandeur, their scenery compares unfavourably with that of the Andes, while their beauty is surpassed by that of the Alps. Not having seen the Andes, I am unable to criticise the assertion regarding the grandeur of the Himalayas, but I find it difficult to imagine anything finer than their scenery.

As regards beauty, the Himalayas at their best surpass the Alps, because they exhibit far more variety, and present everything on a grander scale.

The Himalayas are a kind of Dr. Jekyll and Mr. Hyde. They have two faces—the fair and the plain. In May they are at their worst. Those of the hillsides which are not afforested are brown, arid, and desolate, and the valleys, in addition to being unpleasantly hot, are dry and dusty. The foliage of the trees lacks freshness, and everywhere there is a remarkable absence of water, save in the valleys through which the rivers flow. On the other hand, September is the month in which the Himalayas attain perfection or something approaching it. The eye is refreshed by the bright emerald garment which the hills have newly donned. The foliage is green and luxuriant. Waterfalls, cascades, mighty torrents and rivulets abound. Himachal has been converted into fairyland by the monsoon rains.

A remarkable feature of the Himalayas is the abruptness with which they rise from the plains in most places. In some parts there are low foothills; but speaking generally the mountains that rise from the plain attain a height of 4000 or 5000 feet.

It is difficult for any person who has not passed from the plains of India to the Himalayas to realise fully the vast difference between the two countries and the dramatic suddenness with which the change takes place.

The plains are as flat as the proverbial pancake—a dead monotony of cultivated alluvium, square mile upon square mile of wheat, rice, vetch, sugar-cane, and other crops, amidst which mango groves, bamboo clumps, palms, and hamlets are scattered promiscuously. In some places the hills rise sheer from this, in others they are separated from the alluvial plains by belts of country known as the Tarai and Bhabar. The Tarai is low-lying, marshy land covered with tall, feathery grass, beautifully monotonous. This is succeeded by a stretch of gently-rising ground, 10 or 20 miles in breadth, known as the Bhabar—a strip of forest composed mainly of tall evergreen *sal* trees (*Shorea robusta*). These trees grow so close together that the forest is difficult to penetrate, especially after the rains, when the undergrowth is dense and rank. Very beautiful is the Bhabar, and very stimulating to the imagination. One writer speaks of it as "a jungle rhapsody, an extravagant, impossible botanical *tour de force*, intensely modern in its Titanic, incoherent magnificence." It is the home of the elephant, the tiger, the panther, the wild boar, several species of deer, and of many strange and beautiful birds.

Whether from the flat plains or the gently-sloping Bhabar, the mountains rise with startling suddenness.

The flora and fauna of the Himalayas differ from those of the neighbouring plains as greatly as the trees and animals of England differ from those of Africa.

Of the common trees of the plains of India—the *nim*, mango, babul, tamarind, shesham, palm, and plantain—not one is to be found growing on the hills. The lower slopes are covered with *sal* trees like the Bhabar. These cease to grow at elevations of 3000 feet above the sea-level, and, higher up, every rise of 1000 feet means a considerable change in the flora. Above the *sal* belt come several species of tropical evergreen trees, among the stems and branches of which great creepers entangle themselves in fantastic figures. At elevations of 4000 feet the long-leaved pine (*Pinus longifolia*) appears. From 5000 to 10,000 feet, several species of evergreen oaks abound. Above 6000 feet are to be seen the rhododendron, the deodar and other hill cypresses, and the beautiful horse-chestnut. On the lower slopes the undergrowth is composed largely of begonias and berberry. Higher up maidenhair and other ferns abound, and the trunks of the oaks and rhododendrons are festooned with hanging moss.

Between elevations of 10,000 and 12,000 feet the silver fir is the commonest tree. Above 12,000 feet the firs become stunted and dwarfed, on account of the low temperatures that prevail, and juniper and birch are the characteristic trees.

There are spots in the Himalayas, at heights varying from 10,000 to 12,000 feet, where wild raspberries grow, and the yellow colt's-foot, the dandelion, the blue gentian, the Michaelmas daisy, the purple columbine, the centauria, the anemone, and the edelweiss occur in profusion. Orchids grow in large numbers in most parts of the Himalayas.

Every hillside is not covered with foliage. Many are rugged and bare. Some of these are too precipitous to sustain vegetation, others are masses of quartz and granite. On the hillsides most exposed to the wind, only grass and small shrubs are able to obtain a foothold.

"On the vast ridges of elevated mountain masses," writes Weber in *The Forests of Upper India*, "which constitute the Himalayas are found different regions of distinct character. The loftiest peaks of the snowy range abutting on the great plateaux of Central Asia and Tibet run like a great belt across the globe, falling towards the south-west to the plains of India. Between the summit and the plains, a distance of 60 to 70 miles, there are higher, middle, and lower ranges, so cut up by deep and winding valleys and river-courses, that no labyrinth could be found more confusing or difficult to unravel. There is nowhere any tableland, as at the Cape or in Colorado, with horizontal strata of rock cut down by water into valleys or cañons. The strata seem, on the contrary, to have been shoved up and crumpled in all directions by some powerful shrinkage of the earth's crust, due perhaps to cooling; and the result is such a jumble of contorted rock masses, that it looks as if some great castle had been blown up by dynamite and its walls hurled in all directions. The great central masses, however, consist generally of crystalline granite, gneiss, and quartz rock, protruding from the bowels of the earth and shoving up the stratified envelope of rocks nearly 6 miles above sea-level.... The higher you get up ... the rougher and more difficult becomes the climbing; the valleys are deeper and more cut into ravines, the rocks more fantastically and rudely torn asunder, and the very vitals of the earth exposed; while the heights above tower to the skies. The torrents rushing from under the glaciers which flow from the snow-clad summits roar and foam, eating their way ever into the misty gorges."

Those who have not visited the Himalayas may perhaps best obtain an idea of the nature of the country from a brief description of that traversed by a path leading from the plain to the snowy range. Let us take the path from Kathgodam, the terminus of the Rohilkhand and Kumaun railway, to the Pindari glacier.

For the first two miles the journey is along the cart-road to Naini Tal, on the right bank of the Gola river.

At Ranibagh the pilgrim to the Pindari glacier leaves the cart-road and follows a bridle-path which, having crossed the Gola by a suspension bridge, mounts the steep hill on the left bank. Skirting this hill on its upward course, the road reaches the far side, which slopes down to the Barakheri stream. A fairly steep ascent of 5 miles through well-wooded country brings the traveller to Bhim Tal, a lake 4500 feet above the level of the sea. This lake, of which the area is about 150 acres, is one of the largest of a series of lakes formed by the flow of mountain streams into cup-like valleys. The path skirts the lake and then ascends the Gagar range, which attains a height of over 7000 feet. From the pass over this range a very fine view is obtainable. To the north the snowy range stretches, and between it and the pass lie 60 miles of mountain and valley. To the south are to be seen Bhim Tal, Sat Tal, and other lakes, nestling in the outer ranges, and, beyond the hills, the vast expanse of the plains.

The Gagar range is well wooded. The majority of the trees are rhododendrons: these, when they put forth their blossoms in spring, display a mass of crimson colouring. From the Gagar pass the road descends for some 3 miles through forest to the valley of the Ramganga. For about a mile the path follows the left bank of this small stream; it then crosses it by a suspension bridge, and forthwith begins to mount gradually the bare rocky Pathargarhi mountain. On the mountain side, a few hundred feet above the Ramganga, is a village of three score double-storeyed houses. These are very picturesque. Their white walls are set off by dark brown woodwork. But alas they are as whited sepulchres. It is only from a distance that they are picturesque. They are typical abodes of the hill folk.

From the Pathargarhi pass the path makes a steep descent down a well-wooded mountain-side to the Deodar stream. After crossing this by a stone bridge, the path continues its switch-back course upwards on a wooded hillside to the Laldana Binaik pass, whence it descends gradually for 6 miles, through first rhododendron then pine forest to the Sual river. This river is crossed by a suspension bridge. From the Sual the path makes an ascent of 3 miles on a rocky hillside to Almora, which is 36 miles from Kathgodam.

Almora used to be a Gurkha stronghold, and is now a charming little hill station situated some 5300 feet above the sea-level.

The town and the civil and military station are built on a saddle-backed ridge which is about 2 miles in length.

The Almora hill was almost completely denuded of trees by the Gurkhas, but the ridge has since become well wooded. Deodar, pine, *tun*, horse-chestnut, and alder trees are plentiful, and throughout the cantonment grows a spiræa hedge.

The avifauna of Almora is very interesting, consisting as it does of a strange mixture of hills and plains birds. Among the latter the most prominent are the grey-necked crow, the koel, the myna, the king-crow and the magpie-robin. In the spring paradise flycatchers are very abundant.

From Almora the road to the snowy range runs over an almost treeless rocky mountain called Kalimat, which rises to a height of 6500 feet. From Kalimat the road descends to Takula—16 miles from Almora. Then there is a further descent of 11 miles to Bageswar—a small town situated on the Sarju river. The inhabitants of Bageswar lead a sleepy existence for 360 days in the year, awakening for a short time in January, when a big fair is held, to which flock men of Dhanpur, Thibetans, Bhotias, Nepalese, Garwalis, and Kumaunis. These bring wool, borax, and skins, which they exchange for the produce of the plains.

From Bageswar the Pindari road is almost level for 22 miles, and runs alongside the Sarju. At first the valley is wide and well cultivated. Here and there are studded villages, of which the houses are roofed with thatching composed of pine needles.

At a place about 16 miles above Bageswar the valley of the Sarju suddenly contracts into a gorge with precipitous cliffs.

The scenery here is superb. The path passes through a shady glade in the midst of which rushes the roaring, foaming river. The trunks and larger branches of the trees are covered with ferns and hanging moss. The landscape might well be the original for a phase of a transformation scene at a pantomime. In the midst of this glade the stream is crossed by a wooden bridge.

At a spot 2 miles above this the path, leaving the Sarju, takes a sharp turn to the left, and begins a steep ascent of 5 miles up the Dhakuri mountain. The base of this hill is well wooded. Higher up the trees are less numerous. On the ridge the rhododendron and oak forest alternates with large patches of grassland, on which wild raspberries and brightly-coloured alpine flowers grow.

From the summit of the Dhakuri mountain a magnificent panorama delights the eye. To the north is a deep valley, above which the snow-clad mountains rise almost precipitously. Towering above the observer are the peaks of the highest mountains in British territory. The peaks and 14,000 feet of the slopes are covered with snow. Below the snow is a series of glaciers: these are succeeded by rocks, grass, and stunted vegetation until the tree-line is reached.

To the south lies the world displayed. Near at hand are 50 miles of rugged mountainous country, and beyond the apparently limitless plains. On a clear day it is said to be possible to distinguish the minarets of Delhi, 300 miles away. In the early morning, when the clouds still hover in the valleys, one seems to gaze upon a white billowy sea studded with rocky islets.

From the Dhakuri pass the path descends about 2000 feet, and then follows the valley of the Pindari river. The scenery here is magnificent. Unlike that of the Sarju, this valley is narrow. It is not much cultivated; amaranthus is almost the only crop grown. The villages are few and the huts which constitute them are rudely constructed. The cliffs are very high, and rise almost perpendicularly, like giant walls, so that the numerous feeders of the river take the form of cascades, in many of which the water falls without interruption for a distance of over 1000 feet.

The Kuphini river joins the Pindar 8 miles from its source. Beyond the junction the path to the glacier crosses to the left bank of the Pindar, and then the ascent becomes steep. During the ascent the character of the flora changes. Trees become fewer and flowers more numerous; yellow colt's-foot, dandelions, gentians, Michaelmas daisies, columbines, centaurias, anemones, and edelweiss grow in profusion. Choughs, monal pheasants, and snow-pigeons are the characteristic birds of this region.

Thus the birds of the Himalayas inhabit a country in every respect unlike the plains of India. They dwell in a different environment, are subjected to a different climate, and feed upon different food. It is therefore not surprising that the two avifaunas should exhibit great divergence. Nevertheless few people who have not actually been in both localities are able to realise the startlingly abrupt transformation of the bird-fauna seen by one who passes from the plains to the hills.

The 5-mile journey from Rajpur to Mussoorie transports the traveller from one bird-realm to another.

The caw of the house-crow is replaced by the deeper note of the corby. Instead of the crescendo shriek of the koel, the pleasing double note of the European cuckoo meets the ear. For the eternal *coo-coo-coo-coo* of the little brown dove, the melodious *kokla-kokla* of the hill green-pigeon is substituted. The harsh cries of the rose-ringed paroquets give place to the softer call of the slaty-headed species. The monotonous *tonk-tonk-tonk* of the coppersmith and the *kutur-kutur-kutur* of the green barbet are no more heard; in their stead the curious calls of the great Himalayan barbet resound among the hills. The dissonant voices of the seven sisters no longer issue from the thicket; their place is taken by the weird but less unpleasant calls of the Himalayan streaked laughing-thrushes. Even the sounds of the night are different. The chuckles and cackles of the spotted owlets no longer fill the welkin; the silence of the darkness is broken in the mountains by the low monotonous whistle of the pigmy-collared owlet.

The eye equally with the ear testifies to the traveller that when he has reached an altitude of 5000 feet he has entered another avian realm. The golden-backed woodpecker, the green bee-eater, the "blue jay" or roller, the paddy bird, the Indian and the magpie-robin, most familiar birds of the plains, are no longer seen. Their places are taken by the blue-magpies, the beautiful verditer flycatcher, the Himalayan and the black-headed jays, the black bulbul, and tits of several species.

All the birds, it is true, are not new. Some of our familiar friends of the plains are still with us. There are the kite, the scavenger vulture, the common myna, and a number of others, but these are the exceptions which prove the rule.

Scientific ornithologists recognise this great difference between the two faunas, and include the Himalayas in the Palæarctic region, while the plains form part of the Oriental region.

The chief things which affect the distribution of birds appear to be food-supply and temperature. Hence it is evident that in the Himalayas the avifauna along the snow-line differs greatly from that of the low, warm valleys. The range of temperature in all parts of the hills varies greatly with the season. At the ordinary hill stations the minimum temperature in the summer is sometimes as high as 70°, while in the winter it may drop to 23° F. Thus in midwinter many of the birds which normally live near the snow-line at 12,000 feet descend to 7000 or 6000 feet, and not a few hill birds leave the Himalayas for a time and tarry in the plains until the severity of the winter has passed away.

THE COMMON BIRDS OF THE WESTERN HIMALAYAS

THE CORVIDÆ OR CROW FAMILY

This family, which is well represented in the Himalayas, includes the true crows, with their allies, the choughs, pies, jays, and tits.

The common Indian house-crow (*Corvus splendens*), with which every Anglo-Indian is only too familiar, loveth not great altitudes, hence does not occur in any of the higher hill stations. Almora is the one place in the hills where he appears to be common. There he displays all the shameless impudence of his brethren in the plains.

The common crow of the Himalayas is the large all-black species which is known as the Indian corby or jungle crow (*C. macrorhynchus*). Unlike its grey-necked cousin, this bird is not a public nuisance; nevertheless it occasionally renders itself objectionable by carrying off a chicken or a tame pigeon. In May or June it constructs, high up in a tree, a rough nest, which is usually well concealed by the thick foliage. The nest is a shallow cup or platform in the midst of which is a depression, lined with grass and hair. Horse-hair is used in preference to other kinds of hair; if this be not available crows will use human hair, or hair plucked from off the backs of cattle. Those who put out skins to dry are warned that nesting crows are apt to damage them seriously. Three or four eggs are laid. These are dull green, speckled with brown. Crows affect great secrecy regarding their nests. If a pair think that their nursery is being looked at by a human being, they show their displeasure by swearing as only crows can, and by tearing pieces of moss off the branch of some tree and dropping these on the offender's head!

Two species of chough, the red-billed (*Graculus eremita*), which is identical with the European form, and the yellow-billed chough (*Pyrrhocorax alpinus*), are found in the Himalayas; but he who would see them must either ascend nearly to the snow-line or remain on in the hills during the winter.

Blue-magpies are truly magnificent birds, being in appearance not unlike small pheasants. Two species grace the Himalayas: the red-billed (*Urocissa occipitalis*) and the yellow-billed blue-magpie (*U. flavirostris*). These are distinguishable one from the other mainly by the colour of the beak. A blue-magpie is a bird over 2 feet in length, of which the fine tail accounts for three-fourths. The head, neck, and breast are black, and the remainder of the plumage is a beautiful blue with handsome white markings. It is quite unnecessary to describe the blue-magpie in detail. It is impossible to mistake it. Even a blind man cannot fail to notice it because of its loud ringing call. East of Simla the red-billed species is by far the commoner, while to the west the yellow-billed form rules the roost. The vernacular names for the blue-magpie are *Nilkhant* at Mussoorie and *Dig-dall* at Simla.

The Himalayan tree-pie (*Dendrocitta himalayensis*), although a fine bird, looks mean in comparison with his blue cousins. This species is like a dull edition of the tree-pie of the plains. It is dressed like a quaker. It is easily recognised when on the wing. Its flight is very characteristic, consisting of a few rapid flaps of the pinions followed by a sail on outstretched wings. The median pair of tail feathers is much longer than the others, the pair next to the middle one is the second longest, and the outer one shortest of all. Thus the tail, when expanded during flight, has a curious appearance.

We now come to the jays. That brilliant study in light and dark blue, so common in the plains, which we call the blue-jay, does not occur in the Himalayas; nor is it a jay at all: its proper name is the Indian roller (*Coracias indica*). It is in no way connected with the jay tribe, being not even a passerine bird. We know this because of the arrangement of its deep plantar tendons, because its palate is desmognathous instead of ægithognathous, because— but I think I will not proceed further with these reasons; if I do, this article will resemble a letter written by the conscientious undergraduate who used to copy into each of his epistles to his mother, a page of *A Complete Guide to the Town of Cambridge.* The fond mother doubtless found her son's letters very instructive, but they were not exactly what she wanted. Let it suffice that the familiar bird with wings of two shades of blue is not a jay, nor even one of the Corviniæ, but a blood relation of the kingfishers and bee-eaters.

Two true jays, however, are common in the Western Himalayas. These are known to science as the Himalayan jay (*Garrulus bispecularis*) and the black-throated jay (*G. lanceolatus*). The former is a fawn-coloured bird, with a black moustachial streak. As birds do not usually indulge in moustaches, this streak renders the bird an easy one to identify. The tail is black, and the wing has the characteristic blue band with narrow black cross-bars. This species goes about in large noisy flocks. Once at Naini Tal I came upon a flock which cannot have numbered fewer than forty individuals.

The handsome black-throated jay is a bird that must be familiar to every one who visits a Himalayan hill station with his eyes open. Nevertheless no one seems to have taken the trouble to write about it. Those who have compiled lists of birds usually dismiss it in their notes with such adjectives as "abundant," and "very common." It is remarkable that many popular writers should have discoursed upon the feathered folk of the plains, while few have devoted themselves to the interesting birds of the hills. There seem to be two reasons for this neglect of the latter. Firstly, it is only the favoured few to whom it is given to spend more than ten days at a time in the cool heights; most of us have to toil in the hot plains. Secondly, the thick foliage of the mountain-side makes bird-watching a somewhat difficult operation. The observer frequently catches sight of an interesting-looking bird, only to see it disappear among the foliage before he has had time even to identify it.

The black-throated jay is a handsome bird, more striking in appearance even than the jay of England (*G. glandarius*). Its crested head is black. Its back is a beautiful French grey, its wings are black and white with a bar of the peculiar shade of blue which is characteristic of the jay family and so rarely seen in nature or art. Across this blue bar run thin black transverse lines. The tail is of the same blue with similar black cross-bars, and each feather is tipped with white. The throat is black, with short white lines on it. The legs are pinkish slaty, and the bill is slate coloured in some individuals, and almost white in others. The size of this jay is the same as that of our familiar English one. Black-throated jays go about in flocks. This is a characteristic of a great many Himalayan birds. Probably the majority of the common birds of these mountains lead a sociable existence, like that of the "seven sisters" of the plains. A man may walk for half-an-hour through a Himalayan wood without seeing a bird or hearing any bird-sound save the distant scream of a kite or the raucous voice of the black crow; then suddenly he comes upon quite a congregation of birds, a flock of a hundred or more noisy laughing-thrushes, or numbers of cheeping white-eyes and tits, or it may be a flock of rowdy black bulbuls. All the birds of the wood seem to be collected in one place. This flocking of the birds in the hills must, I think, be accounted for by the fact that birds are by nature sociable creatures, and that food is particularly abundant. In a dense wood every tree offers either insect or vegetable food, so that a large number of birds can live in company without fear of starving each other out. In the plains food is less abundant, hence most birds that dwell there are able to gratify their fondness for each other's society only at roosting time; during the day they are obliged to separate, in order to find the wherewithal to feed upon.

Like all sociable birds, the black-throated jay is very noisy. Birds have a language of a kind, a language composed entirely of interjections, a language in which only the simplest emotions—fear, joy, hunger, and maternal care—can be expressed. Now, when a considerable flock of birds is wandering through a dense forest, it is obvious that the individuals which compose it would be very liable to lose touch with one another had they no means of informing one another of their whereabouts. The result is that such a means has been developed. Every bird, whose habit it is to go about in company, has the habit of continually uttering some kind of call or cry. It probably does this unconsciously, without being aware that it is making any sound.

In Madras a white-headed babbler nestling was once brought to me. I took charge of it and fed it, and noticed that when it was not asleep it kept up a continuous cheeping all day long, even when it was eating, although it had no companion. The habit of continually uttering its note was inherited. When the flock is stationary the note is a comparatively low one; but when an individual makes up its mind to fly any distance, say ten or a dozen yards, it gives vent to a louder call, so as to inform its companions that it is moving. This sound seems to induce others to follow its lead. This is especially noticeable in the case of the white-throated laughing-thrush. I have seen one of these birds fly to a branch in a tree, uttering its curious call, and then hop on to another branch in the same tree. Scarcely has it left the first branch when a second laughing-thrush flies to it; then a fourth, a fifth, and so on; so that the birds look as though they might be playing "Follow the man from Cook's." The black-throated jay is noisy even for a sociable bird. The sound which it seems to produce more often than any other is very like the harsh anger-cry of the common myna. Many Himalayan birds have rather discordant notes, and in this respect these mountains do not compare favourably with the Nilgiris, where the blithe notes of the bulbuls are very pleasing to the ear.

Jays are by nature bold birds. They are inclined to be timid in England, because they are so much persecuted by the game-keeper. In the Himalayas they are as bold as the crow. It is not uncommon to see two or three jays hopping about outside a kitchen picking up the scraps pitched out by the cook. Sometimes two jays make a dash at the same morsel. Then a tiff ensues, but it is mostly made up of menacing screeches. One bird bears away the coveted morsel, swearing lustily, and the unsuccessful claimant lets him go in peace. When a jay comes upon a morsel of food too large to be swallowed whole, it flies with it to a tree and holds it under one foot and tears it up with its beak. This is a characteristically corvine habit. The black-throated jay is an exceedingly restless bird; it is always on the move. Like its English cousin, it is not a bird of very powerful flight. As Gilbert White says: "Magpies and jays flutter with powerless wings, and make no despatch." In the Himalayas there is no necessity for it to make much despatch; it rarely has to cover any distance on the wing. When it does fly a dozen yards or so, its passage is marked by much noisy flapping of the pinions.

The nutcrackers can scarcely be numbered among the common birds, but are sometimes seen in our hill stations, and, such is the "cussedness" of birds that if I omit to notice the nutcrackers several are certain to show themselves to many of those who read these lines. A chocolate-brown bird, bigger than a crow, and spotted and barred with white all over, can be nothing other than one of the Himalayan nutcrackers. It may be the

Himalayan species (*Nucifraga hemispila*), or the larger spotted nutcracker (*N. multipunctata*).

The members of the crow family which I have attempted to describe above are all large birds, birds bigger than a crow. It now behoves us to consider the smaller members of the corvine clan.

The tits form a sub-family of the crows. Now at first sight the crow and the tit seem to have but little in common. However, close inspection, whether by the anatomist or the naturalist, reveals the mark of the corvidæ in the tits. First, there is the habit of holding food under the foot while it is being devoured. Then there is the aggressiveness of the tits. This is Lloyd-Georgian or even Winstonian in its magnitude. "Tits," writes Jerdon, "are excessively bold and even ferocious, the larger ones occasionally destroying young and sickly birds, both in a wild state and in confinement."

Many species of tit dwell in the Himalayas. To describe them all would bewilder the reader; I will, therefore, content myself with brief descriptions of four species, each of which is to be seen daily in every hill station of the Western Himalayas.

The green-backed tit (*Parus monticola*) is a glorified edition of our English great tit. It is a bird considerably smaller than a sparrow.

The cheeks are white, the rest of the head is black, as are the breast and a characteristic line running along the abdomen. The back is greenish yellow, the lower parts are deep yellow. The wings are black with two white bars, the tail is black tipped with white. This is one of the commonest birds in most hill stations.

Like the sparrow, it is ever ready to rear up its brood in a hole in the wall of a house. Any kind of a hole will do, provided the aperture is too small to admit of the entrance of birds larger than itself.

The nesting operations of a pair of green-backed tits form the subject of a separate essay.

Another tit much in evidence is the yellow-cheeked tit, *Machlolophus xanthogenys*. I apologise for its scientific name. Take a green-backed tit, paint its cheeks bright yellow, and give it a black crest tipped with yellow, and you will have transformed him into a yellow-cheeked tit.

There remain to be described two pigmy tits. The first of these is that feathered exquisite, the red-headed tit (*Ægithaliscus erythrocephalus*). I will not again apologise for the name; it must suffice that the average ornithologist is never happy unless he be either saddling a small bird with a big name or altering the denomination of some unfortunate fowl. This fussy little mite is not so long as a man's thumb. It is crestless; the spot where the crest ought to be is chestnut red. The remainder of the upper plumage is bluish grey, while the lower plumage is the colour of rust. The black face is set off by a white eyebrow. Last, but not least, of our common tits is the crested black tit (*Lophophanes melanopterus*). The crested head and breast of this midget are black. The cheeks and nape are white, while the rest of the upper plumage is iron grey.

There is yet another tit of which mention must be made, because he is the common tit of Almora. The climate of Almora is so much milder than that of other hill stations that its birds are intermediate between those of the hills and the plains. The Indian grey tit (*Parus atriceps*) is a bird of wide distribution. It is the common tit of the Nilgiris, is found in many of the better-wooded parts of the plains, and ascends the Himalayas up to 6000 feet. It is a grey bird with the head, neck, breast, and abdominal line black. The cheeks are white. It is less gregarious than the other tits. Its notes are harsh and varied, being usually a *ti-ti-chee* or *pretty-pretty*.

I have not noticed this species at either Mussoorie or Naini Tal, but, as I have stated, it is common at Almora.

As has been mentioned above, tits usually go about in flocks. It is no uncommon thing for a flock to contain all of the four species of tit just described, a number of white-eyes, some nuthatches, warblers, tree-creepers, a woodpecker or two, and possibly some sibias and laughing-thrushes.

THE CRATEROPODIDÆ OR BABBLER FAMILY

The Crateropodidæ form a most heterogeneous collection of birds, including, as they do, such divers fowls as babblers, whistling-thrushes, bulbuls, and white-eyes. Whenever a systematist comes across an Asiatic bird of which he can make nothing, he classes it among the Crateropodidæ. This is convenient for the systematist, but embarrassing for the naturalist.

The most characteristic members of the family are those ugly, untidy, noisy earth-coloured birds which occur everywhere in the plains, and always go about in little companies, whence their popular name "seven sisters."

To men of science these birds are known as babblers. Babblers proper are essentially birds of the plains. In the hills they are replaced by their cousins, the laughing-thrushes. Laughing-thrushes are merely glorified babblers. The Himalayan streaked laughing-thrush (*Trochalopterum lineatum*) is one of the commonest of the birds of our hill stations. It is a reddish brown fowl, about eight inches long. Each of its feathers has a black shaft; it is these dark shafts that give the bird its streaked appearance. Its chin, throat, and breast are chestnut-red, and on each cheek there is a patch of similar hue. The general appearance of the streaked laughing-thrush is that of one of the seven sisters who is wearing her best frock. Like their sisters of the plains, Himalayan streaked laughing-thrushes go about in small flocks and are exceedingly noisy. Sometimes a number of them assemble, apparently for the sole purpose of holding a speaking competition. They are never so happy as when thus engaged.

Streaked laughing-thrushes frequent gardens, and, as they are inordinately fond of hearing their own voices, it is certainly not their fault if they escape observation. By way of a nest they build a rough-and-ready cup-shaped structure in a low bush or on the ground; but, as Hume remarked, "the bird, as a rule, conceals the nest so well that, though a loose, and for the size of the architect, a large structure, it is difficult to find, even when one closely examines the bush in which it is."

Three other species of laughing-thrush must be numbered among common birds of the Himalayas, although they, like the heroine of *A Bad Girl's Diary*, are often heard and not seen. The white-throated laughing-thrush (*Garrulax albigularis*) is a handsome bird larger than a myna. Its general colour is rich olive brown. It has a black eyebrow and shows a fine expanse of white shirt front. It goes about in large flocks and continually utters a cry, loud and plaintive and not in the least like laughter.

The remaining laughing-thrushes are known as the rufous-chinned (*Ianthocincla rufigularis*) and the red-headed (*Trochalopterum erythrocephalum*). The former may be distinguished from the white-throated species by the fact that the lower part only of its throat is white, the chin being red. The red-headed laughing-thrush has no white at all in the under parts. The next member of the family of the Crateropodidæ that demands our attention is the rusty-cheeked scimitar-babbler (*Pomatorhinus erythrogenys*).

Scimitar-babblers are so called because of the long, slender, compressed beak, which is curved downwards like that of a sunbird.

Several species of scimitar-babbler occur in the Himalayas. The above mentioned is the most abundant in the Western Himalayas. This species is known as the *Banbakra* at Mussoorie. Its bill is 1½ inch long. The upper plumage is olive brown. The forehead, cheeks, sides of the neck, and thighs are chestnut-red, as is a patch under the tail. The chin and throat and the median portion of the breast and abdomen are white with faint grey stripes. Scimitar-babblers have habits similar to those of laughing-thrushes. They go about in pairs, seeking for insects among fallen leaves. The call is a loud whistle.

Very different in habits and appearance from any of the babblers mentioned above is the famous Himalayan whistling-thrush (*Myiophoneous temmincki*). To see this bird it is necessary to repair to some mountain stream. It is always in evidence in the neighbourhood of the dhobi's ghat at Naini Tal, and is particularly abundant on the banks of the Kosi river round about Khairna. At first sight the Himalayan whistling-thrush looks very like a cock blackbird. His yellow bill adds to the similitude. It is only when he is seen with the sun shining upon him that the cobalt blue patches in his plumage are noticed. His habit is to perch on the boulders which are washed by the foaming waters of a mountain torrent. On these he finds plenty of insects and snails, which constitute the chief items on his menu. He pursues the elusive insect in much the same way as a wagtail does, calling his wings to his assistance when chasing a particularly nimble creature. He has the habit of frequently expanding his tail. This species utters a loud and pleasant call, also a shrill cry like that of the spotted forktail. All torrent-haunting birds are in the habit of uttering such a note; indeed it is no easy task to distinguish between the alarm notes of the various species that frequent mountain streams.

Of very different habits is the black-headed sibia (*Lioptila capistrata*). This species is strictly arboreal. As mentioned previously, it is often found in company with flocks of tits and other gregarious birds. It feeds on insects, which it picks off the leaves of trees. Its usual call is a harsh twitter. It is a reddish brown bird, rather larger than a bulbul, with a black-crested head. There is a white bar on the wing.

The Indian white-eye (*Zosterops palbebrosa*) is not at all like any of the babblers hitherto described. In size, appearance, and habits, it approximates closely to the tits, with which it often consorts. Indeed, Jerdon calls the bird the white-eyed tit. It occurs in all well-wooded parts of the country, both in the plains and the hills. No bird is easier to identify. The upper parts are greenish yellow, and the lower bright yellow, while round the eye runs a broad conspicuous ring of white feathers, whence the popular names of the species, white-eye and spectacle-bird. Except at the breeding season, it goes about in flocks of considerable size. Each individual utters unceasingly a low, plaintive, sonorous, cheeping note. As was stated above, all arboreal gregarious birds have this habit. It is by means of this call note that they keep each other apprised of their whereabouts. But for such a signal it would scarcely be possible for the flock to hold together. At the breeding season the cock white-eye acquires an unusually sweet song. The nest is an exquisite little cup, which hangs, like a hammock, suspended from a slender forked branch. Two pretty pale blue eggs are laid.

A very diminutive member of the babbler clan is the fire-cap (*Cephalopyrus flammiceps*). The upper parts of its plumage are olive green; the lower portions are golden yellow. In the cock the chin is suffused with red. The cock wears a further ornament in the shape of a cap of flaming red, which renders his identification easy.

Until recently all ornithologists agreed that the curious starling-like bird known as the spotted-wing (*Psaroglossa spiloptera*) was a kind of aberrant starling, but systematists have lately relegated it to the Crateropodidæ. At Mussoorie the natives call it the *Puli*. Its upper parts are dark grey spotted with black. The wings are glossy greenish black with white spots. The lower parts are reddish. A flock of half-a-dozen or more birds having a starling-like appearance, which twitter like stares and keep to the topmost branches of trees, may be set down safely as spotted-wings.

We now come to the last of the Crateropodidæ—the bulbuls. These birds are so different from most of their brethren that they are held to constitute a sub-family. I presume that every reader is familiar with the common bulbul of the plains. To every one who is not, my advice is that he should go into the verandah in the spring and look among the leaves of the croton plants. The chances are in favour of this search leading to the discovery of a neat cup-shaped nest owned by a pair of handsome crested birds, which wear a bright crimson patch under the tail, and give forth at frequent intervals tinkling notes that are blithe and gay.

Both the species of bulbul common in the plains ascend the lower ranges of the Himalayas. These are the Bengal red-vented bulbul (*Molpastes bengalensis*) and the Bengal red-whiskered bulbul (*Otocompsa emeria*).

The addition of the adjective "Bengal" is important, for every province of India has its own special species of bulbul.

The Molpastes bulbul is a bird about half as big again as the sparrow, but with a longer tail. The black head is marked by a short crest. The cheeks are brown. There is a conspicuous crimson patch under the tail. The remainder of the plumage is brown, but each feather on the body is margined with creamy white, so that the bird is marked by a pattern that is, as "Eha" pointed out, not unlike the scales on a fish. Both ends of the tail feathers are creamy white.

Otocompsa is a far more showy bird. The crest is long and pointed and curves forward a little over the bill. There is the usual crimson patch under the tail and another on each cheek. The rest of the cheek is white, as is the lower plumage. A black necklace, interrupted in front, marks the junction of the throat and the breast. Neither of these bulbuls ascends the hills very high, but I have seen the former at the Brewery below Naini Tal.

The common bulbul of the Himalayas is the white-cheeked species (*Molpastes leucogenys*). This bird, which is very common at Almora, has the habits of its brethren in the plains. Its crest is pointed and its cheeks are white like those of an Otocompsa bulbul. But it has rather a weedy appearance and lacks the red feathers on the sides of the head. The patch of feathers under the tail is bright sulphur-yellow instead of crimson.

The only other species of bulbul commonly seen in the hills is a very different bird. It is known as the black bulbul (*Hypsipetes psaroides*).

The bulbuls that we have been considering are inoffensive little birds which lead quiet and respectable lives. Not so the black bulbuls. These are aggressive, disreputable-looking creatures which go about in disorderly, rowdy gangs.

The song of most bulbuls is a medley of pleasant tinkling notes; the cries of the black bulbuls are harsh and unlovely.

Black bulbuls look black only when seen from a distance. When closely inspected their plumage is seen to be dark grey. The bill and legs are red. The crest, I regret to say, usually looks the worse for wear. Black bulbuls seem never to descend to the ground. They keep almost exclusively to tops of lofty trees. They are very partial to the nectar enclosed within the calyces of rhododendron flowers. A party of half a dozen untidy black birds, with moderately long tails, which keep to the tops of trees and make much noise, may with certainty be set down as black bulbuls.

These curious birds form the subject of a separate essay.

THE SITTIDÆ OR NUTHATCH FAMILY

The Sittidæ are a well-defined family of little birds. When not occupied with domestic cares, they congregate in small flocks that run up and down the trunks and branches of trees in search of insects. The nuthatch most commonly seen in the hills is the white-tailed species (*Sitta himalayensis*). The general hue of this bird is slaty blue. The forehead and a broad line running down the sides of the head and neck are black. There is a good deal of white in the tail, which is short in this and in all species of nuthatch. The underparts are of a chestnut hue. The Himalayan nuthatch is very partial to the red berries of *Arisæma jacque-montii*—a small plant of the family to which the arums and the "lords and ladies" belong. Half a dozen nuthatches attacking one of the red spikes of this plant present a pretty sight. The berries ripen in July and August, and at Naini Tal one rarely comes across a complete spike because the nuthatches pounce upon every berry the moment it is ripe.

THE DICRURIDÆ OR DRONGO FAMILY

The famous black drongo or king-crow (*Dicrurus ater*) is the type of this well-marked family of passerine birds. The king-crow is about the size of a bulbul, but he has a tail 6 or 7 inches long, which is gracefully forked. His whole plumage is glossy jet black. He loves to sit on a telegraph wire or other exposed perch, and thence make sallies into the air after flying insects. He is one of the commonest birds in India. His cheery call—half-squeak, half-whistle—must be familiar to every Anglo-Indian. As to his character, I will repeat what I have said elsewhere: "The king-crow is the Black Prince of the bird world—the embodiment of pluck. The thing in feathers of which he is afraid has yet to be evolved. Like the mediæval knight, he goes about seeking those on whom he can perform some small feat of arms. In certain parts of India he is known as the kotwal—the official who stands forth to the poor as the impersonation of the might and majesty of the British raj."

The king-crow is fairly abundant in the hills. On the lower ranges, and especially at Almora, it is nearly as common as in the plains. On the higher slopes, however, it is largely replaced by the ashy drongo (*Dicrurus longicaudatus*). At most hill stations both species occur. The note of the ashy drongo differs considerably from that of the king-crow: otherwise the habits of the two species are very similar. Take thirty-three per cent. off the pugnacity of the king-crow and you will arrive at a fair estimate of that of the ashy drongo. The latter looks like a king-crow with an unusually long tail, a king-crow of which the black plumage has worn grey like an old broadcloth coat.

The handsome *Bhimraj* or larger racket-tailed drongo (*Dissemurus paradiseus*), a glorified king-crow with a tail fully 20 inches in length, is a Himalayan bird, but he dwells far from the madding crowd, and is not likely to be seen at any hill station except as a captive.

THE CERTHIIDÆ OR WREN FAMILY

The only member of this family common about our hill stations is the Himalayan tree-creeper (*Certhia himalayana*). This is a small brown bird, striped and barred with black, which spends the day creeping over the trunks of trees seeking its insect quarry. It is an unobtrusive creature, and, as its plumage assimilates very closely to the bark over which it crawls, it would escape observation more often than it does, but for its call, which is a shrill one.

THE SYLVIIDÆ OR WARBLER FAMILY

The sylviidæ comprise a large number of birds of small size and, with a few exceptions, of plain plumage. The result is that the great majority of them resemble one another so closely that it is as difficult to identify them when at large as it is to see through a brick wall. Small wonder, then, that field naturalists fight rather shy of this family. Of the 110 species of warbler which exist in India, I propose to deal with only one, and that favoured bird is Hodgson's grey-headed flycatcher-warbler (*Cryptolopha xanthoschista*). My reasons for raising this particular species from among the vulgar herd of warblers are two. The first is that it is the commonest bird in our hill stations. The second is that it is distinctively coloured, and in consequence easy to identify.

It is impossible for a human being to visit any hill station between Murree and Naini Tal in spring without remarking this warbler. I do not exaggerate when I say that its voice issues from every second tree.

This species may be said to be *the* warbler of the Western Himalayas, and, as such, it has been made the subject of a separate essay.

THE LANIIDÆ OR SHRIKE FAMILY

The butcher-birds are the best-known members of this fraternity. Undoubtedly passerine in structure, shrikes are as indubitably raptores by nature. They are nothing less than pocket hawks.

Their habit is to sit on an exposed perch and pounce from thence on to some insect on the ground. The larger species attack small birds.

Four species of butcher-bird may perhaps be classed among the common birds of the Himalayas; but they are inhabitants of the lower ranges only. It is unusual to see a shrike at as high an elevation as 6000 feet. In consequence they are seldom observed at hill stations.

It is true that the grey-backed shrike does occur as high as 9000 feet, but this species, being confined mainly to the inner ranges, does not occur at most hill stations.

The bay-backed shrike (*Lanius vittatus*) is a bird rather smaller than a bulbul. Its head is grey except for a broad black band running through the eye. The wings and tail are black and white. The back is chestnut red and the rump white.

The rufous-backed shrike (*L. erythronotus*) is very like the last species, but it is a larger bird. It has no white in the wings and tail, and its rump is red instead of being white.

The grey-backed shrike (*L. tephronotus*) is very like the rufous-backed species, but may be distinguished by the fact that the grey of the head extends more than half-way down the back.

As its name indicates, the black-headed shrike (*L. nigriceps*) has the whole head black; but the cheeks, chin, and throat are white.

Butcher-birds are of striking rather than beautiful appearance. They have some very handsome relatives which are known as minivets. Every person must have seen a company of small birds with somewhat long tails, clothed in bright scarlet and black—birds which flit about among the trees like sparks driven before the wind. These are cock minivets. The hens, which are often found in company with them, are in their way equally beautiful and conspicuous, for they are bright yellow in those parts of the plumage where the cocks are scarlet. It is impossible to mistake a minivet, but it is quite another matter to say to which species any particular minivet belongs. The species commonly seen about our hill stations are *Pericrocotus speciosus*, the Indian scarlet minivet, and *P. brevirostris*, the short-billed minivet. The former is 9 inches long, while the latter is but 7½. Again, the red of the former is scarlet and that of the latter crimson rather than scarlet. These distinctions are sufficiently apparent when two species are seen side by side, but are scarcely sufficient to enable the ordinary observer to determine the species of a flock seen flitting about amid the foliage. This, however, need not disturb us. Most people are quite satisfied to know that these exquisite little birds are all called minivets.

THE ORIOLIDÆ OR ORIOLE FAMILY

The beautiful orioles are birds of the plains rather than of the hills. One species, however, the Indian Oriole (*Oriolus kundoo*) is a summer visitor to the Himalayas. The cock is a bright yellow bird with a pink bill. There is some black on his cheeks and wing feathers. The hen is less brilliantly coloured, the yellow of her plumage being dull and mixed with green. Orioles are a little larger than bulbuls. They rarely, if ever, descend to the ground. I do not remember having seen the birds at Murree, Mussoorie, or Naini Tal, but they are common at Almora in summer.

THE STURNIDÆ OR STARLING FAMILY

The Himalayan starling (*Sturnus humii*) is so like his European brother in appearance that it is scarcely possible to distinguish between the two species unless they are seen side by side. Is it necessary to describe the starling? Does an Englishman exist who is not well acquainted with the vivacious bird which makes itself at home in his garden or on his housetop in England? We have all admired its dark plumage, which displays a green or bronze sheen in the sunlight, and which is so curiously spotted with buff.

The Himalayan species is, I think, common only in the more westerly parts of the hills.

The common myna (*Acridotheres tristis*) is nearly as abundant in the hills as it is in the plains. I should not have deemed it necessary to describe this bird, had not a lady asked me a few days ago whether a pair of mynas, which were fighting as only mynas can fight, were seven sisters.

The myna is a bird considerably smaller than a crow. His head, neck, and upper breast are black, while the rest of his plumage is quaker brown, save for a broad white wing-bar, very conspicuous during flight, and some white in the tail. The legs and bill look as though they had been dipped in the mustard pot, and there is a bare patch of mustard-coloured skin on either side of the head. This sprightly bird is sociably inclined. Grasshoppers form its favourite food. These it seeks on the grass, over which it struts with as much dignity as a stout raja. In the spring the mynas make free with our bungalows, seizing on any convenient holes or ledges as sites for their nests. The nest is a conglomeration of straw, rags, paper, and any rubbish that comes to beak. The eggs are a beautiful blue.

The only other myna commonly seen in Himalayan hill stations is the jungle myna (*Æthiopsar fuscus*). This is so like the species just described, that nine out of ten people fail to differentiate between the two birds. Close inspection shows that this species has a little tuft of feathers on the forehead, which the common myna lacks. On the other hand, the yellow patch of skin round the eyes is wanting in the jungle myna.

THE MUSCICAPIDÆ OR FLYCATCHER FAMILY

The family of the flycatchers is well represented in the hills, for its members love trees. The great majority of them seem never to descend to the ground at all. Flycatchers are birds that feed exclusively on insects, which they catch on the wing. Their habit is to make from some perch little sallies into the air after their quarry. But, we must bear in mind that a bird that behaves thus is not necessarily a flycatcher. Other birds, as, for example, king-crows and bee-eaters, have discovered how excellent a way this is of securing a good supply of food. The beautiful verditer flycatcher (*Stoparola melanops*) must be familiar to everyone who has visited the Himalayas. The plumage of this flycatcher is pale blue—blue of that peculiar shade known as verditer blue. There is a little black on the head. The plumage of the hen is distinctly duller than that of the cock. This species loves to sit on a telegraph wire or at the very summit of a tree and pour forth its song, which consists of a pleasant, if somewhat harsh, trill or warble of a dozen or more notes. The next flycatcher that demands notice is the white-browed blue flycatcher (*Cyornis superciliaris*). In this species the hen differs considerably from the cock in appearance. The upper plumage of the latter is a dull blue, set off by a white eyebrow. The lower plumage is white save for a blue collaret, which is interrupted in the middle. The upper plumage of the hen is olive brown, washed with blue in parts. Beneath she is pale buff. This species, like the last, nests in a hole.

There are yet four other species of flycatcher which, although less frequently seen than the two just mentioned, deserve place among the common birds of the Himalayas. Two of these are homely-looking little creatures, while two are as striking as it is possible for a fowl of the air to be, and this is saying a great deal.

The brown flycatcher (*Alseonax latirostris*) is a bird that may pass for a small sparrow if not carefully looked at. Of course its habits are very different to those of the sparrow; moreover, it has a narrow ring of white feathers round the eye. The grey-headed flycatcher (*Culicicapa ceylonensis*) is a species of which the sexes are alike. The head, neck, and breast are grey; the wings and tail are brown; the back is dull yellow, and the lower plumage bright yellow. Notwithstanding all this yellow, the bird is not conspicuous except during flight, because the wings when closed cover up nearly all the yellow. This bird frequents all the hill streams. At Naini Tal any person may be tolerably certain of coming across it by going down the Khairna road to the place where that road meets the stream. The nest of this species is a beautiful pocket of moss attached to some moss-covered rock or tree.

The rufous-bellied niltava (*Niltava sundara*) or fairy blue-chat, as Jerdon calls it, is the kind of bird one would expect to find in fairyland. The front and sides of the head, and the chin and throat of the cock are deep velvety black. His crown, nape, and lower back, and a spot on cheeks and wings, are glistening blue. He also sports some light blue in his tail. His lower plumage is chestnut red. The upper plumage of the hen is olive brown save for a brilliant blue patch on either side of the head. Her tail is chestnut red. This beautiful species is about the size of a sparrow.

Even more splendid is the paradise flycatcher (*Terpsiphone paradisi*). The hen, and the cock, when he is quite young, look rather like specimens of the bulbul family, being rich chestnut-hued birds with the head and crest metallic bluish black. The hen is content with a gown of this style throughout her life. Not so the cock. No sooner does he reach the years of discretion than he assumes a magnificent caudal appendage. His two middle tail feathers suddenly begin to grow, and go on growing till they become three or four times as long as he is, and so flutter behind him in the wind like streamers when he flies. Nor does he rest content with this finery. When he is about three years old he doffs his chestnut plumage, and in its place dons a snowy white one. He is then a truly magnificent object. The first time one catches sight of this white bird with his satin streamers floating behind him, one wonders whether he is but an object seen in a dream.

This flycatcher is a regular visitor in summer to Almora, where it nests. Six thousand feet appear to be about the limit of its ascent, and in consequence this beautiful creature is not common at any of the higher hill stations. I have seen it at the brewery below Naini Tal, but not at Naini Tal itself.

THE TURDIDÆ OR THRUSH FAMILY

This large family is well represented in the hills, and embraces a number of beautiful and interesting birds.

The dark grey bush-chat (*Oreicola ferrea*) is as common in the hills as is the robin in the plains. It is about the size of a robin. The upper plumage of the cock is grey in winter and black in summer. This change in colour is the result of wear and tear suffered by the feathers. Each bird is given by nature a new suit of clothes every autumn, and in most cases the bird, like a Government *chaprassi*, has to make it last a whole year. Both eat, drink, sleep, and do everything in their coats. There is, however, this difference between the bird and the *chaprassi*: the plumage of the former always looks clean and smart, while the garment of the *chaprassi* is usually neither the one nor the other. The coat of the dark grey bush-chat is made up of black feathers edged with grey. As the margins of the feathers alone show, the bird looks grey so long as the grey margins exist, and when these wear away it appears black. The cock has a conspicuous white eyebrow, and displays some white in his wings and tail. He is quite a dandy. The hen is a reddish brown bird with a pale grey eyebrow. This species likes to pretend it is a flycatcher. The flycatchers proper do not object in the least; in this country of multitudinous insects there are more than enough for every kind of bird.

Brief mention must be made here of the Indian bush-chat (*Pratincola maura*), because this chat is common at Almora, and breeds there. I have not seen it at other hill stations. It does not appear to ascend the Himalayas higher than 5500 feet. In the cock the upper parts are black (brown in winter) with a large white patch on each side of the neck. The breast is orange-red. The lower parts are ruddy brown. The hen is a plain reddish brown bird.

We now come to what is, in my opinion, one of the most striking birds in the Himalayas. I refer to the bird known to men of science as *Henicurus maculatus*, or the western spotted forktail. Those Europeans who are not men of science call it the hill-wagtail on account of its habits, or the *dhobi* bird because of its unaccountable predilection for the spot where the grunting, perspiring washerman pursues his destructive calling. The head and neck of this showy bird are jet black save for a conspicuous white patch running from the centre of the crown to the base of the bill, which gives the bird a curious appearance. The shoulders are decorated by a cape or tippet of black, copiously spotted with white. The wings are black and white. The tail feathers are black, but each has a broad white band at the tip, and, as the two median feathers are the shortest, and each succeeding pair longer, the tail has, when closed, the appearance of being composed of alternate broad black and narrow white V-shaped bars. The lower back and rump are white, but these are scarcely visible except during flight or when the bird is preening its feathers. The legs are pinkish white. This forktail is a trifle larger than a wagtail, and its tail is over 6 inches in length. It is never found away from streams.

I will not dilate further upon the habits of this bird because a separate essay is devoted to it.

Two other water-birds must now be mentioned. These love not the *dhobi*, and dwell by preference far from the madding crowd. They are very common in the interior of the hills, and everyone who has travelled in the inner ranges must be familiar with them, even if he do not know what to call them. The white-capped redstart (*Chimarrhornis leucocephalus*) is a bird that compels attention. His black plumage looks as though it were made of rich velvet. On his head he wears a cap as white as snow. His tail, rump, and abdomen are bright chestnut red, so that, as he leaps into the air after the circling gnat, he looks almost as if he were on fire.

The third common bird of Himalayan streams is the plumbeous redstart or water-robin (*Rhyacornis fuliginosus*). This species is very robin-like in appearance. The body is dusky indigo blue; the tail and abdomen are ferruginous. The habits of this and the bird just described are similar. Both species love to disport themselves on rocks and boulders lapped by the gentle-flowing stream in the valley, or lashed by the torrent on the hillside. Like all redstarts, these constantly flirt the tail.

The grey-winged ouzel (*Merula boulboul*) is perhaps the finest songster in the Himalayas. Throughout the early summer the cock makes the wooded hillsides ring with his blackbird-like melody. The grey-winged ouzel is a near relative of the English blackbird. Take a cock blackbird and paint his wings dark grey, and cover his bill with red colouring matter, and you will have to all appearances a grey-winged ouzel. In order to effect the transformation of the brown female, it is only necessary to redden her bill.

The nesting operations of this species are described in the essay near the end of Part I.

Two other species allied to the grey-winged ouzel demand our attention. The first is the blue-headed rock-thrush (*Petrophila cinclorhyncha*). This is not like any bird found in England. The head, chin, and throat of the cock are cobalt blue; there is also a patch of this colour on his wing; the sides of the head and neck are black, as are the back and wing feathers. The rump and lower parts are chestnut. The hen, as is the case with many of her sex, is an inconspicuous olive-brown bird. This species spends most of its time on the ground, and frequents, as its name implies, open rocky ground.

The last of the Turdidæ which has to be considered is the small-billed mountain-thrush (*Oreocincla dauma*). This bird is as like the thrush of our English gardens as one pea is like another. Unfortunately it does not visit gardens in this country, and is not a very common bird.

THE FRINGILLIDÆ OR FINCH FAMILY

The vulgar sparrow and the immaculate canary are members of this large and flourishing family of birds. The distinguishing feature of the finches is a massive beak, admirably adapted to the husking of the grain on which the members of the family feed largely. In some species, as for example the grosbeaks, the bill is immensely thick. Only one species of grosbeak appears to be common in the Himalayas. This is *Pycnorhamphus icteroides*, the black-and-yellow grosbeak. The colouring of the cock is so like that of the black-headed oriole that it is doubtless frequently mistaken for the latter.

This bird forms the subject of a separate essay, where it is fully described.

The Himalayan greenfinch (*Hypacanthis spinoides*) is an unobtrusive little bird that loves to sit at the summit of a tree and utter a forlorn *peee* fifty times a minute. It is a dull green bird with some yellow on the head, neck, and back; the abdomen is of a brighter hue of yellow.

The house-sparrow, like the house-crow, is a bird of the plains rather than of the hills. The common sparrow of the Himalayas is the handsome cinnamon tree-sparrow (*Passer cinamomeus*). The cock is easily recognised by his bright cinnamon-coloured head and shoulders. Imagine a house-sparrow shorn of sixty per cent. of his impudence, and you will have arrived at a fair estimate of the character of the tree-sparrow.

The only other members of the Finch family that concern us are the buntings. A bunting is a rather superior kind of sparrow—a Lord Curzon among sparrows—a sparrow with a refined beak. The familiar English yellowhammer is a bunting. Two buntings are common in the Western Himalayas. The first of these, the eastern meadow-bunting (*Emberiza stracheyi*), looks like a large, well-groomed sparrow. A broad slate-coloured band runs from the base of the beak over the top of the head to the nape of the neck. In addition to this, there are on each side of the head blackish bars, like those on the head of the quail. By these signs the bird may be recognised. The other species is the white-capped bunting (*Emberiza stewarti*). This is a chestnut-coloured bird with a pale grey cap. Buntings associate in small flocks and affect open rather than well-wooded country. They are not very interesting birds.

THE HIRUNDINIDÆ OR SWALLOW FAMILY

A small bird that spends hours together on the wing, dashing through the air at great speed, frequently changing its course, now flying high, now just skimming the ground, must be either a swallow or a swift. Many people are totally at a loss to distinguish between a swallow and a swift. The two birds differ anatomically. A swift is not a passerine bird. It cannot perch. When it wants to take a rest it has to repair to its nest. Swallows, on the other hand, are fond of settling on telegraph wires. It is quite easy to distinguish between the birds when they are on the wing. A flying swift may be compared to an anchor with enormous flukes (the wings), or to an arrow (the body) attached to a bow (the wings). As the swift dashes through the air at a speed of fully 100 miles an hour, it never closes its wings to the sides of its body; it merely whips the air rapidly with the tips of them. On the other hand, the swallow, when it flies, closes its wings to its body at every stroke. Notwithstanding its greater effort, it does not move nearly so rapidly as the swift. The swifts will be considered in their proper place. Three species of swallow are likely to be seen in the Himalayas. A small ashy brown swallow with a short tail is the crag-martin (*Ptyonoprogne rupestris*).

The common swallow of England (*Hirundo rustica*) occurs in large numbers at all hill stations in the Himalayas. This bird should require no description. Its glossy purple-blue plumage, the patches of chestnut red on the forehead and throat, and the elegantly-forked tail must be familiar to every Englishman. As in England, this bird constructs under the eaves of roofs its nest of mud lined with feathers.

Not unlike the common swallow, but readily distinguishable from it in that the lower back is chestnut red, is *Hirundo nepalensis*—Hodgson's striated swallow, or the red-rumped swallow, as Jerdon well called it. This bird also breeds under eaves. Numbers of red-rumped swallows are to be seen daily seeking their insect quarry over the lake at Naini Tal.

THE MOTACILLIDÆ OR WAGTAIL FAMILY

The great majority of the wagtails are merely winter visitors to India. Thus they are likely to be seen in the hills only when resting from their travels. That is to say, in April and May, when homeward bound, or in September and October, when they move southwards. A few wagtails, however, tarry in the hills till quite late in the season. The wagtail most likely to be seen is the grey wagtail (*Motacilla melanope*). This species, notwithstanding its name, has bright yellow lower plumage. It nests in Kashmir.

Allied to the wagtails are the pipits. These display the elegant form of the wagtail and the sober colouring of the lark.

They affect open country and feed on the ground. The upland pipit (*Oreocorys sylvanus*) is the common species of the Himalayas. It constructs a nest of grass on the ground, into which the common cuckoo, of which more anon, frequently drops an egg.

THE NECTARINIDÆ OR SUNBIRD FAMILY

The sunbirds are feathered exquisites. They take in the Old World the place in the New World occupied by the humming-birds. Sunbirds, however, are superior to humming-birds in that they possess the gift of song. They are not particularly abundant in the Himalayas, and, as they do not seem to occur west of Garhwal, I am perhaps not justified in giving them a place in this essay.

I do so because one species is fairly common round about Naini Tal. I have seen this bird—the Himalayan yellow-backed sunbird (*Æthopyga scheriæ*)—flitting about, sucking honey from the flowers in the verandah of the hotel at the brewery below Naini Tal.

The head and neck of the cock are glistening green. The back, shoulders, chin, throat, breast, and sides of the head are crimson.

The lower parts are greenish yellow. The two median tail feathers are longer than the others. The bill is long and curved. The hen is a comparatively dull greenish-brown bird.

THE DICÆIDÆ OR FLOWER-PECKER FAMILY

The fire-breasted flower-pecker (*Dicæum ignipectus*) is perhaps the smallest bird in India. Its total length does not exceed 3 inches. The upper parts are greenish black and the lower parts buff. The cock has a large patch of crimson on his breast, with a black patch lower down. As this species frequents lofty trees, it is usually seen from below, and the crimson breast renders the cock unmistakeable.

THE PICIDÆ OR WOODPECKER FAMILY

Woodpeckers abound in the well-wooded Himalayas.

The woodpecker most commonly seen in the western hill stations is the brown-fronted pied species (*Dendrocopus auriceps*). This is a black bird, spotted and barred with white: some might call it a white bird, heavily spotted and barred with black. The forehead is amber brown. That is the distinguishing feature of this species. The cock has a red-and-gold crest, which the hen lacks. Both sexes rejoice in a crimson patch under the tail—a feature common to all species of pied woodpecker. *Dendrocopus auriceps* nests earlier in the year than do most hill-birds, so that by the time the majority of the European visitors arrive in the hills, the young woodpeckers have left their nest, which is a hole excavated by the parents in a tree, a rhododendron by preference.

Two other species of pied woodpecker are common in the hills—the rufous-bellied (*Hypopicus hypererythrus*) and the Western Himalayan species (*Dendrocopus himalayensis*). The former is particularly abundant at Murree. These two species are distinguished from the brown-fronted pied woodpecker by having no brown on the forehead. The rufous abdomen serves to differentiate the rufous-bellied from the Western Himalayan species. The above woodpeckers are not much larger than mynas.

There remains yet another common species—the West Himalayan scaly-bellied green woodpecker (*Gecinus squamatus*). The English name of this bird is very cumbrous. There is no help for this. Numerous adjectives and adjectival adjuncts are necessary to each species to distinguish it from each of the host of other woodpeckers. This particular species is larger than a crow and is recognisable by its green colour. It might be possible to condense an accurate description of the plumage of this bird into half a column of print. I will, however, refrain. There is a limit to the patience of even the Anglo-Indian.

THE CAPITONIDÆ OR BARBET FAMILY

The only member of this family common in the Himalayas is that fine bird known as the great Himalayan barbet (*Megalæma marshallorum*). As this forms the subject of a separate essay, detailed description is unnecessary in the present one. It will suffice that the bird is over a foot in length and has a large yellow beak. Its prevailing hue is grass green. It has a bright red patch under the tail. It goes about in small flocks and constantly utters a loud plaintive dissyllabic note.

THE ALCEDINIDÆ OR KINGFISHER FAMILY

The Himalayan pied kingfisher (*Ceryle lugubris*) is a bird as large as a crow. Its plumage is speckled black and white, like that of a Hamburg fowl. It feeds entirely on fish, and frequents the larger hill streams. Its habit is to squat on a branch, or if the day be cloudy, on a boulder in mid-stream, whence it dives into the water after its quarry. Sometimes, kestrel-like, it hovers in the air on rapidly-vibrating pinions until it espies a fish in the water below, when it closes its wings and drops with a splash in the water, to emerge with a silvery object in its bill.

THE UPUPIDÆ OR HOOPOE FAMILY

The unique hoopoe (*Upupa epops*) next demands our attention. This is a bird about the size of a myna. The wings and tail are boldly marked with alternate bands of black and white. The remainder of the plumage is of a fawn colour. The bill is long and slender, like that of a snipe, but slightly curved. The crest is the feature that distinguishes the hoopoe from all other birds. This opens and closes like a lady's fan. Normally it remains closed, but when the bird is startled, and at the moment when the hoopoe alights on the ground, the crest opens to form a magnificent corona. Hoopoes seek their food on grass-covered land, digging insects out of the earth with their long, pick-like bills. They are very partial to a dust-bath. During the breeding season—that is to say, in April and May in the Himalayas—hoopoes continually utter in low tones *uk-uk-uk*. The call is not unlike that of the coppersmith, but less metallic and much more subdued. The flight of the hoopoe is undulating or jerky, like that of a butterfly. Young hoopoes are reared up in a hole in a building, or in a bank. The nest is incredibly malodoriferous.

THE CYPSELIDÆ OR SWIFT FAMILY

The flight and general appearance of the swifts have already been described. The common Indian swift (*Cypselus affinis*) is perhaps the bird most frequently seen in the Himalayas. A small dark sooty brown bird with a broad white bar across the back, a living monoplane that dashes through the air at the rate of 100 miles an hour, continually giving vent to what Jerdon has so well described as a "shivering scream," can be none other than this species. It nests under the eaves of houses or in verandahs. Hundreds of these swifts nest in the Landour bazar, and there is scarcely a *dak* bungalow or a deserted building in the whole of Kumaun which does not afford nesting sites for at least a dozen pairs of swifts. About sunset these birds indulge in riotous exercise, dashing with loud screams in and out among the pillars that support the roof of the verandah in which their nests are placed. The nest is composed of mud and feathers and straw. The saliva of the swift is sticky and makes excellent cement.

The other swift commonly seen in the Himalayas is the Alpine swift (*Cypselus melba*). This is distinguishable from the Indian species by its white abdomen and dark rump. It is perhaps the swiftest flier among birds. Like the species already described, it utters a shrill cry when on the wing.

THE CUCULIDÆ OR CUCKOO FAMILY

It is not possible for anyone of sound hearing to be an hour in a hill station in the early summer without being aware of the presence of cuckoos. The Himalayas literally teem with them. From March to June, or even July, the cheerful double note of the common cuckoo (*Cuculus canorus*) emanates from every second tree. This species, as all the world knows, looks like a hawk and flies like a hawk.

According to some naturalists, the cuckoo profits by its similarity to a bird of prey. The little birds which it imposes upon are supposed to fly away in terror when they see it, thus allowing it to work unmolested its wicked will in their nests. My experience is that little birds have a habit of attacking birds of prey that venture near their nest. The presence of eggs or young ones makes the most timid creatures as bold as the proverbial lion. I therefore do not believe that these cuckoos which resemble birds of prey derive any benefit therefrom.

The hen European cuckoo differs very slightly from the cock. In some species, as, for example, the famous "brain-fever bird" (*Hierococcyx varius*), there is no external difference between the sexes, while in others, such as the Indian koel (*Eudynamis honorata*), and the violet cuckoo (*Chrysococcyx xanthorhynchus*), the sexes are very dissimilar. I commend these facts to the notice of those who profess to explain sexual dimorphism (the different appearance of the sexes) by means of natural or sexual selection. The comfortable theory that the hens are less showily coloured than the cocks, because they stand in greater need of protective colouring while sitting on the nest, cannot be applied to the parasitic cuckoos, for these build no nests, neither do they incubate their eggs.

In the Himalayas the common cuckoo victimises chiefly pipits, larks, and chats, but its eggs have been found in the nests of many other birds, including the magpie-robin, white-cheeked bulbul, spotted forktail, rufous-backed shrike, and the jungle babbler.

The eggs of *Cuculus canorus* display considerable variation in colour. Those who are interested in the subject are referred to Mr. Stuart Baker's papers on the Oology of the Indian Cuckoos in Volume XVII of the *Journal of the Bombay Natural History Society*.

It often happens that the eggs laid by the cuckoo are not unlike those of the birds in the nests of which they are deposited. Hence, some naturalists assert that the cuckoo, having laid an egg, flies about with it in her bill until she comes upon a clutch which matches her egg. Perhaps the best reply to this theory is that such refinement on the part of the cuckoo is wholly unnecessary. Most birds, when seized by the mania of incubation, will sit upon anything which even remotely resembles an egg.

Mr. Stuart Baker writes that he has not found that there is any proof of the cuckoo trying to match its eggs with those of the intended foster-mother, or that it selects a foster-mother whose eggs shall match its own. He adds that not one of his correspondents has advanced this suggestion, and states that he has little doubt that convenience of site and propinquity to the cuckoo about to lay its eggs are the main requisitions.

Almost indistinguishable from the common cuckoo in appearance is the Himalayan cuckoo (*Cuculus saturatus*). The call of this bird, which continues later in the year than that of the common cuckoo, is not unlike the *whoot-whoot-whoot* of the crow-pheasant or coucal. Perhaps it is even more like the *uk-uk-uk* of the hoopoe repeated very loudly. It may be syllabised as *cuck-hoo-hoo-hoo-hoo*. Not very much is known about the habits of this species. It is believed to victimise chiefly willow-warblers.

The Indian cuckoo (*Cuculus micropterus*) resembles in appearance the two species already described. Blanford speaks of its call as a fine melodious whistle. I would not describe the note as a whistle. To me it sounds like *wherefore, wherefore*, impressively and sonorously intoned. The vernacular names *Boukotako* and *Kyphulpakka* are onomatopoetic, as is Broken Pekoe Bird, by which name the species is known to many Europeans.

Last, but not least of the common Himalayan cuckoos, are the famous brain-fever birds, whose crescendo *brain-fever, BRAIN-FEVER, BRAIN-FEVER*, which is shrieked at all hours of the day and the night, has called forth untold volumes of awful profanity from jaded Europeans living in the plains, and has earned the highest encomiums of Indians.

There are two species of brain-fever bird that disport themselves in the Himalayas. These are known respectively as the large and the common hawk-cuckoo (*Hierococcyx sparverioides* and *H. varius*). I do not profess to distinguish with certainty between the notes of these two birds, but am under the impression that the larger form is the one that makes itself heard at Naini Tal and Mussoorie.

The Indian koel (*Eudynamis honorata*) is not to be numbered among the common birds of the Himalayas. Its noisy call *kuil, kuil, kuil*, which may be expressed by the words *you're-ill, you're-ill, who-are-you? who-are-you?* is heard throughout the sub-Himalayan regions in the early summer, and I have heard it as high up as Rajpur below Mussoorie, but have not noticed the bird at any of the hill stations except Almora. As has already been stated, the avifauna of Almora, a little station in the inner hills nearly forty miles from the plains, is a very curious one. I have not only heard the koel calling there, but have seen a young koel being fed by crows. Now, at Almora alone of the hill stations does *Corvus splendens*, the Indian house-crow, occur, and this is the usual victim of the koel. I would therefore attribute the presence of the koel at Almora and its absence from other hill stations to the fact that at Almora alone the koel's dupe occurs.

THE PSITTACIDÆ OR PARROT FAMILY

The parrots are not strongly represented in the Himalayas. Only one species is commonly seen at the various hill stations. This is the slaty-headed paroquet (*Palæornis schisticeps*). In appearance it closely resembles the common green parrot of the plains (*P. torquatus*), differing chiefly in having the head slate coloured instead of green. The cock, moreover, has a red patch on the shoulder. The habits of the slaty-headed paroquet are those of the common green parrot: its cries, however, are less harsh, and it is less aggressively bold. The pretty little western blossom-headed paroquet (*P. cyanocephalus*) ascends the hills to a height of some 5000 feet. It is recognisable by the fact that the head of the cock is red, tinged with blue like the bloom on a plum.

THE STRIGIDÆ OR OWL FAMILY

We now come to those much-abused birds—the owls. The Himalayas, in common with most other parts of the world, are well stocked with these pirates of the night. The vast majority of owls, being strictly nocturnal, escape observation. Usually the presence of any species of owl in a locality is made known only by its voice. I may here remark that diurnal birds know as little about nocturnal birds as the man in the street does, hence the savage manner in which they mob any luckless owl that happens to be abroad in the daytime. Birds are intensely conservative; they resent strongly what they regard as an addition to the local avifauna. This assertion may be proved by setting free a cockatoo in the plains of India. Before the bird has been at large for ten minutes it will be surrounded by a mob of reviling crows.

The collared pigmy owlet (*Glaucidium brodiei*) is perhaps the commonest owl in the Himalayas: at any rate, it is the species that makes itself heard most often. Those who sit out of doors after dinner cannot fail to have remarked a soft low whistle heard at regular intervals of about thirty seconds. That is the call of the pigmy collared owlet. The owlet itself is a tiny creature, about the size of a sparrow. Like several other little owls, it sometimes shows itself during the daytime. Once at Mussoorie I noticed a pigmy collared owlet sitting as bold as brass on a conspicuous branch about midday and making grimaces at me. The other species likely to be heard at hill stations are the brown wood-owl (*Syrnium indrani*), the call of which has been syllabised *to-whoo*, and the little spotted Himalayan scops owl (*Scops spilocephalus*), of which the note is double whistle *who-who*.

THE VULTURIDÆ OR VULTURE FAMILY

From the owls to the diurnal birds of prey it is but a short step. Next to the warblers, the raptores are the most difficult birds to distinguish one from the other. Nearly all of them are creatures of mottled-brown plumage, and, as the plumage changes with the period of life, it is impossible to differentiate them by descriptions of their colouring.

The vultures are perhaps the ugliest of all birds. Most of them have the head devoid of feathers, and they are thus enabled to bury this member in their loathsome food without soiling their feathers. In the air, owing to the magnificent ease with which they fly, they are splendid objects. Their habit is to rise high above the earth and hang motionless in the atmosphere on outstretched wings, or sail in circles without any perceptible motion of the pinions. Vultures are not the only raptorial birds that do this. Kites are almost equally skilled. But kites are distinguished by having a fairly long tail, that of vultures being short and wedge shaped. The sides of the wings of the vultures are straight, and the wings stand out at right angles to the body. In all species, except the scavenger vulture, the tips of the wings are turned up as the birds float or sail in the air, and the ends of the wings are much cut up, looking like fingers.

Perhaps the commonest vulture of the Himalayas is that very familiar fowl—the small white scavenger vulture (*Neophron ginginianus*), often called Pharaoh's chicken and other opprobrious names that I will not mention. This bird eats everything that is filthy and unclean. The natural consequence is that it looks untidy and disreputable. It is, without exception, the ugliest bird in the world. It is about the size of a kite. The plumage is a dirty white, except the edges of the wing feathers, which are shabby black. The naked face is of a pale mustard colour, as are the bill and legs. The feathers on the back of the head project like the back hairs of an untidy schoolboy. Its walk is an ungainly waddle. Nevertheless—so great is the magic of wings—this bird, as it soars high above the earth, looks a noble fowl; it then appears to be snow-white with black margins to the wings.

Another vulture frequently met with is the Indian white-backed vulture (*Pseudogyps bengalensis*). The plumage of this species is a very dark grey, almost black. The naked head is rather lighter than the rest of the body. The lower back is white: this makes the bird easy to identify when it is perched. It has some white in the wings, and this, during flight, is visible as a very broad band that runs from the body nearly to the tip of the wing. Thus the wing from below appears to be white with broad black edges. During flight this species may be distinguished from the last by the fingered tips of its wings, by both edges of the wing being black and the body being dark instead of white.

The third common vulture is the Himalayan griffon (*Gyps himalayensis*). This is distinguishable from the two species already described by having no white in the wings.

The lammergeyer or bearded vulture (*Gypætus barbatus*) is the king of the vultures. Some ornithologists classify it with the eagles. It is a connecting link between the two families. It is 4 feet in length and is known to the hillmen as the Argul.

During flight it may be recognised by the whitish head and nape, the pale brown lower plumage and the dark rounded tail.

Usually it keeps to rocky hills and mountains, over which it beats with a steady, sailing, vulturine flight. Numerous stories are told of its swooping down and carrying off young children, lambs, goats, and other small animals. Those who will may believe these stories. I do not. The lammergeyer is quite content to make a meal of offal, old bones, or other refuse.

THE FALCONIDÆ OR FAMILY OF BIRDS OF PREY

First and foremost of the Falconidæ are the eagles. Let me preface what little I have to say about these birds with the remark that I am unable to set forth any characteristics whereby a novice may recognise an eagle when he sees one on the wing. The reader should disabuse his mind of the idea he may have obtained from the writings of the poets of the grandeur of the eagle. Eagles may be, and doubtless often are, mistaken for kites. They are simply rather large falcons. They are mostly coloured very like the kite.

All true eagles have the leg feathered to the toe. I give this method of diagnosis for what it is worth, and that is, I fear, not very much, because eagles as a rule do not willingly afford the observer an opportunity of inspecting their tarsi.

The eagles most commonly seen in the Himalayas are the imperial eagle (*Aquila helica*), the booted eagle (*Hieraëtus pennatus*), Bonelli's eagle (*Hieraëtus fasciatus*), the changeable hawk-eagle (*Spizaëtus limnaëtus*), and Hodgson's hawk-eagle (*Spizaëtus nepalensis*).

The imperial eagle has perhaps the darkest plumage of all the eagles. This species does not live up to its name. It feeds largely on carrion, and probably never catches anything larger than a rat. The imperial eagle is common about Mussoorie except in the rains. Captain Hutton states that he has seen as many as fifty of them together in the month of October when they reassemble after the monsoon.

The booted eagle has a very shrill call. Its lower parts are pale in hue.

Bonelli's eagle is fairly common both at Naini Tal and Mussoorie. It is a fine bird, and has plenty of courage. It often stoops to fowls and is destructive to game birds. It is of slighter build than the two eagles above described. Its lower parts are white.

The changeable hawk-eagle is also a fine bird. It is very addicted to peafowl. The hillmen call it the *Mohrhaita*, which, being interpreted, is the peacock-killer. It utters a loud cry, which Thompson renders *whee-whick, whee-whick*. This call is uttered by the bird both when on the wing and at rest. Another cry of this species has been syllabised *toot, toot, toot, toot-twee*.

Hodgson's hawk-eagle is also destructive to game. It emits a shrill musical whistle which can sometimes be heard when the bird is so high as to appear a mere speck against the sky. This species has a narrow crest.

Allied to the true eagles are the serpent-eagles. In these the leg is not feathered to the toe, so they may be said to form a link between the true eagles and the falcons.

One species—the crested serpent-eagle (*Spilornis cheela*)—is common in the Himalayas up to 8000 feet.

This eagle is perhaps the most handsome of the birds of prey. The crest is large and imposing. The upper parts are dark brown, almost black, with a purple or green gloss. The breast and under parts are rich deep brown profusely dotted with white ocelli. On the tail and wings are white bars. The wing bars are very conspicuous during flight. The crested serpent-eagle flies with the wings held very far back, so that it looks, as "Exile" says, like a large butterfly. When flying it constantly utters its shrill, plaintive call composed of two short sharp cries and three prolonged notes, the latter being in a slightly higher key.

Of the remaining birds of prey perhaps only two can fairly be numbered among the common birds of the Himalayas, and both of these are easy to recognise. They are the kite and the kestrel.

The common pariah kite (*Milvus govinda*) is the most familiar raptorial bird in India. Hundreds of kites dwell at every hill-station. They spend the greater part of the day on the wing, either sailing gracefully in circles high overhead or gliding on outstretched pinions over mountain and valley, with head pointing downwards, looking for the refuse on which they feed. To mistake a kite is impossible. Throughout the day it makes the welkin ring with its querulous *chee-hee-hee-hee-hee*. Some kites are larger than others, consequently ornithologists, who are never so happy as when splitting up species, have made a separate species of the larger race. This latter is called *Milvus melanotis*, the large Indian kite. It is common in the hills.

The kestrel (*Tinnunculus alaudarius*) is perhaps the easiest of all the birds of prey to identify. It is a greyish fowl with dull brick-red wings and shoulders. Its flight is very distinctive. It flaps the wings more rapidly than do most of its kind. While beating over the country it checks its flight now and again and hovers on rapidly vibrating wings. It does this when it fancies it has seen a mouse, lizard, or other living thing moving on the ground below. If its surmise proves correct, it drops from above and thus takes its quarry completely by surprise. It is on account of this peculiar habit of hovering in the air that the kestrel is often called the wind-hover in England. Needless to say, the kestrel affects open tracts rather than forest country. One of these birds is usually to be seen engaged in its craft above the bare slope of the hill on which Mussoorie is built. Other places where kestrels are always to be seen are the bare hills round Almora. The nest of this species is usually placed on an inaccessible crag.

THE COLUMBIDÆ OR DOVE FAMILY

The cooing community is not much in evidence in the hills. In the Himalayas doves do not obtrude themselves upon our notice in the way that they do in the plains.

The green-pigeon of the mountains is the kokla (*Sphenocercus sphenurus*), so called on account of its melodious call, *kok-la, kok-la*. In appearance it is very like the green-pigeon of the plains and is equally difficult to distinguish from its leafy surroundings. The bronze-winged dove (*Chalcophaps indica*) I have never observed at any hill-station, but it is abundant in the lower ranges and in the Terai. Every sportsman must be familiar with the bird. Its magnificent bronzed metallic, green plumage renders its identification easy. The commonest dove of the Himalayan hill-stations is the Indian turtle-dove (*Turtur ferago*). Its plumage is of that grey hue which is so characteristic of doves as to be called dove-colour. The turtle-dove has a conspicuous patch of black-and-white feathers on each side of the neck. The only other dove seen in the hills with which it can be confounded is the little brown dove (*T. cambayensis*). The latter is a much smaller bird, and I have not observed it anywhere higher than 4500 feet above the sea-level.

The spotted dove (*T. suratensis*) occurs in small numbers in most parts of the Himalayas up to 7000 feet. It is distinguished by the wing coverts being spotted with rufous and black.

The Indian ring-dove (*T. risorius*) also occurs in the Western Himalayas. It is of a paler hue than the other doves and has no patch of black-and-white feathers on the sides of the neck, but has a black collar, with a narrow white border, round the back of the neck.

One other dove should perhaps be mentioned among the common birds of the Himalayas, namely, the bar-tailed cuckoo-dove (*Macropygia tusalia*). A dove with a long barred tail, of which the feathers are graduated, the median ones being the longest, may be set down as this species.

THE PHASIANIDÆ OR FAMILY OF GAME BIRDS

The Himalayas are the home of many species of gallinaceous birds. In the highest ranges the snow-cocks, the tragopans, the blood-pheasant, and the glorious monaul or Impeyan pheasant abound. The foothills are the happy hunting-grounds of the ancestral cock-a-doodle-doo.

As this book is written with the object of enabling persons staying at the various hill-stations to identify the commoner birds, I do not propose to describe the gallinaceous denizens of the higher ranges or the foothills. In the ranges of moderate elevation, on which all the hill-stations are situated, the kalij, the cheer, and the koklas pheasants are common. Of these three the kalij is the only one likely to be seen in the ordinary course of a walk. The others are not likely to show themselves unless flushed by a dog.

The white-crested kalij-pheasant (*Gennæus albicristatus*) may occasionally be seen in the vicinity of a village.

The bird does not come up to the Englishman's ideal of a pheasant. The bushy tail causes it to look rather like a product of the farmyard. The cock is over two feet in length, the hen is five inches shorter. The plumage of the former is dark brown, tinged with blue, each feather having a pale margin. The rump is white with broad black bars. The hen is uniformly brown, each feather having a narrow buff margin. Both sexes rejoice in a long backwardly-directed crest and a patch of bare crimson skin round each eye. The tail is much shorter and more bushy than that of the English pheasant. The crest is white in the cock and reddish yellow in the hen. Baldwin describes the call of this pheasant as "a sharp *twut, twut, twut*. Sometimes very low, with a pause between each note, then suddenly increasing loudly and excitedly."

The kalij usually affords rather poor sport.

The koklas pheasant (*Pucrasia macrolopha*) is another short-tailed species; but it is more game-like in appearance than the kalij and provides better sport.

It may be distinguished from the kalij by its not having the red patch of skin round the eye. The cock of this species has a curious crest, the middle portion of which is short and of a fawn colour; on each side of this is a long lateral tuft coloured black with a green gloss. The cry of this bird has been syllabised as *kok-kok-pokrass*.

In the cheer-pheasant (*Catreus wellichi*) both sexes have a long crest, like that of the kalij, and a red patch of skin round the eye. The tail of this species, however, is long and attenuated like that of the English pheasant, measuring nearly two feet. Wilson says, of the call of this bird: "Both males and females often crow at daybreak and dusk and, in cloudy weather, sometimes during the day. The crow is loud and singular, and, when there is nothing to interrupt, the sound may be heard for at least a mile. It is something like the words *chir-a-pir, chir-a-pir, chir-a-pir, chirwa, chirwa*, but a good deal varied."

The grey quail (*Coturnix communis*) is a common bird of the Himalayas during a few days only in the year. Large numbers of these birds rest in the fields of ripening grain in the course of their long migratory flight. Almost as regularly as clockwork do they appear in the Western Himalayas early in October on their way south, and again in April on their northward journey. By walking through the terraced fields at those times with a gun, considerable bags of quail can be secured. These birds migrate at night. Writing of them, Hume said: "One moonlight night about the third week in April, standing at the top of Benog, a few miles from Mussoorie, a dense cloud many hundred yards in length and fifty yards, I suppose, in breadth of small birds swept over me with the sound of a rushing wind. They were not, I believe, twenty yards above the level of my head, and their quite unmistakable call was uttered by several of those nearest me as they passed."

We must now consider the partridges that patronise the hills. The species most commonly met with in the Himalayas is the chakor (*Caccabis chucar*). In appearance this is very like the French or red-legged partridge, to which it is related. Its prevailing hue is pale reddish brown, the particular shade varying greatly with the individual. The most striking features of this partridge are a black band that runs across the forehead to the eyes and then down the sides of the head round the throat, forming a gorget, and a number of black bars on each flank. The favourite haunts of the chakor are bare grassy hillsides on which a few terraced fields exist. Chakor are noisy birds. The note most commonly heard is the double call from which their name is taken.

The black partridge or common francolin (*Francolinus vulgaris*) is abundant on the lower ranges of the Himalayas. At Mussoorie its curious call is often heard. This is so high-pitched as to be inaudible to some people. To those who can hear it, the call sounds like *juk-juk-tee-tee-tur*. This species has the habit of feigning a broken wing when an enemy approaches its young ones. The cock is a very handsome bird. The prevailing hue of his plumage is black with white spots on the flanks and narrow white bars on the back. The feathers of the crown and wings are buff and dark brown. A chestnut collar runs round the neck, while each side of the head is adorned by a white patch. The whole plumage of the hen is coloured like the wings of the cock.

The common hill-partridge (*Arboricola torqueola*) is a great skulker. He haunts dark densely jungled water-courses and ravines, and so is not likely to be seen about a hill-station; we will therefore pass him over without description.

THE CHARADRIIDÆ OR PLOVER FAMILY

In conclusion mention must be made of the woodcock (*Scolopax rusticola*). This species, although it breeds throughout the Himalayas, usually remains during the summer at altitudes above those at which hill-stations are situate. The lowest height at which its nest has been found is, I believe, 9500 feet.

THE COMMON BIRDS OF THE EASTERN HIMALAYAS

The majority of the birds which are common in the Eastern Himalayas are also abundant in the western part of the range, and have in consequence been described already. In order to avoid repetition this chapter has been put into the form of a list. The list that follows includes all the birds likely to be seen daily by those who in summer visit Darjeeling and other hill-stations east of Nepal.

Of the birds which find place in the list only those are described which have not been mentioned in the essay on the common birds of the Western Himalayas.

Short accounts of all the birds that follow which are not described in this chapter are to be found in the previous one.

THE CORVIDÆ OR CROW FAMILY

1. *Corvus macrorhynchus.* The jungle-crow or Indian corby.

2. *Dendrocitta himalayensis.* The Himalayan tree-pie. Abundant.

3. *Graculus eremita.* The red-billed chough. In summer this species is not usually found much below elevations of 11,000 feet above the sea-level.

4. *Pyrrhocorax alpinus.* The yellow-billed chough. In summer this species is not usually seen at elevations below 11,000 feet.

5. *Garrulus bispecularis.* The Himalayan jay. Not so abundant as in the Western Himalayas.

6. *Parus monticola.* The green-backed tit. A common bird. Very abundant round about Darjeeling.

7. *Machlolophus spilonotus.* The black-spotted yellow tit. This is very like *M. xanthogenys* (the yellow-cheeked tit), which it replaces in the Eastern Himalayas. It is distinguished by having the forehead bright yellow instead of black as in the yellow-cheeked species. It is not very common.

8. *Ægithaliscus erythrocephalus.* The red-headed tit. Very common at Darjeeling.

9. *Parus atriceps.* The Indian grey tit.

THE CRATEROPODIDÆ OR BABBLER FAMILY

Since most species of babblers are notoriously birds of limited distribution, it is not surprising that the kinds common in the Eastern Himalayas should not be the same as those that are abundant west of Nepal.

10. *Garrulax leucolophus.* The Himalayan white-crested laughing-thrush. This is the Eastern counterpart of the white-throated laughing-thrush (*Garrulax albigularis*). This species has a large white crest. It goes about in flocks of about a score. The members of the flock scream and chatter and make discordant sounds which some might deem to resemble laughter.

11. *Ianthocincla ocellata.* The white-spotted laughing-thrush. This is the Eastern counterpart of *Ianthocincla rufigularis*. It has no white in the throat, and the upper plumage is spotted with white. It is found only at high elevations in summer.

12. *Trochalopterum chrysopterum.* The eastern yellow-winged laughing-thrush. This is perhaps the most common bird about Darjeeling. Parties hop about the roads picking up unconsidered trifles.

The forehead is grey, as is much of the remaining plumage. The back of the head is bright chestnut. The throat is chestnut-brown. The wings are chestnut and bright yellow.

13. *Trochalopterum squamatum.* The blue-winged laughing-thrush. This is another common bird. Like all its clan it goes about in flocks. Its wings are chestnut and blue.

14. *Grammatophila striata.* The striated laughing-thrush. A common bird, but as it keeps to dense foliage it is heard more often than seen. Of its curious cries Jerdon likens one to the clucking of a hen which has just laid an egg. The tail is chestnut. The rest of the plumage is umber brown, but every feather has a white streak along the middle. These white streaks give the bird the striated appearance from which it obtains its name.

15. *Pomatorhinus erythrogenys.* The rusty-cheeked scimitar-babbler.

16. *Pomatorhinus schisticeps.* The slaty-headed scimitar-babbler. This is easily distinguished from the foregoing species by its conspicuous white eyebrow.

17. *Alcippe nepalensis.* The Nepal babbler or quaker-thrush. This is a bird smaller than a sparrow. As its popular name indicates, it is clothed in homely brown; but it has a conspicuous ring of white feathers round the eye and a black line on each side of the head, beginning from the eye. It is very

common about Darjeeling. It feeds in trees and bushes, often descending to the ground. It utters a low twittering call.

18. *Stachyrhis nigriceps*. The black-throated babbler or wren-babbler. This is another small bird. Its general hue is olive brown. The throat is black, as is the head, but the latter has white streaks.

It is common about Darjeeling and goes about in flocks that keep to trees.

19. *Stachyrhidopsis ruficeps*. The red-headed babbler or wren-babbler. Another small bird with habits similar to the last.

An olive-brown bird with a chestnut-red cap. The lower parts are reddish yellow.

20. *Myiophoneus temmincki*. The Himalayan whistling-thrush. Common at Darjeeling.

21. *Lioptila capistrata*. The black-headed sibia, one of the most abundant birds about Darjeeling.

22. *Actinodura egertoni*. The rufous bar-wing. A bird about the size of a bulbul. It associates in small flocks which never leave the trees. Common about Darjeeling. A reddish brown bird, with a crest. There is a black bar in the wing.

23. *Zosterops palpebrosa*. The Indian white-eye.

24. *Siva cyanuroptera*. The blue-winged siva or hill-tit. A pretty little bird, about the size of a sparrow. The head is blue, deeper on the sides than on the crown, streaked with brown. The visible portions of the closed wing and tail are cobalt-blue.

This species goes about in flocks and has all the habits of a tit. It utters a cheerful chirrup.

25. *Liothrix lutea*. The red-billed liothrix or hill-tit, or the Pekin-robin. This interesting bird forms the subject of a separate essay.

26. *Ixulus flavicollis*. The yellow-naped ixulus. A small tit-like bird with a crest. Like tits these birds associate in small flocks, which move about amid the foliage uttering a continual twittering.

Brown above, pale yellow below. Chin and throat white. Back of neck rusty yellow. This colour is continued in a demi-collar round the sides of the neck. Common about Darjeeling.

27. *Yuhina gularis*. The striped-throated yuhina. Another tiny bird with all the habits of the tits. A flock of dull-brown birds, about the size of sparrows,

having the chin and throat streaked with black, are likely to be striped-throated yuhinas.

28. *Minla igneitincta.* The red-tailed minla or hill-tit. This tit-like babbler is often seen in company with the true tits, which it resembles in habits and size. The head is black with a white eyebrow. The wings and tail are black and crimson. The rest of the upper plumage is yellowish olive. The throat is white, and the remainder of the lower plumage is bright yellow.

NOTE ON THE TITS AND SMALL BABBLERS

Tits are small birds, smaller than sparrows, which usually go about in flocks. They spend most of their lives in trees. In seeking for insects, on which they feed largely, they often hang upside down from a branch. All tits have these habits; but all birds of these habits are not tits. Thus the following of the babblers described above have all the habits of tits: the white-eye, the black-throated babbler, the red-headed babbler, the blue-winged siva, the yellow-naped ixulus, the striped-throated yuhina, and the red-tailed minla.

The above are all birds of distinctive colouring and may be easily distinguished.

Other small birds which are neither tits nor babblers go about in flocks, as, for example, nuthatches, but these other birds differ in shape and habits from babblers and tits, so that no one is likely to confound them with the smaller Corvidæ or Crateropodidæ.

29. *Molpastes leucogenys.* The white-cheeked bulbul. Common below elevations of 5000 feet.

30. *Hypsipetes psaroides.* The Himalayan black bulbul. Not very common.

31. *Alcurus striatus.* The striated green bulbul. Upper plumage olive-green with yellow streaks. Cheeks dark brown, streaked with pale yellow. Chin and throat yellow, with dark spots on throat. Patch under tail bright yellow.

Striated green bulbuls go about in flocks which keep to the tops of trees. They utter a mellow warbling note. They are abundant about Darjeeling.

THE SITTIDÆ OR NUTHATCH FAMILY

32. *Sitta himalayensis.* Very abundant in the neighbourhood of Darjeeling.

THE DICRURIDÆ OR DRONGO FAMILY

33. *Dicrurus longicaudatus.* The Indian Ashy Drongo.

THE CERTHIIDÆ OR WREN FAMILY

34. *Certhia discolor.* The Sikhim tree-creeper. This species displaces the Himalayan tree-creeper in the Eastern Himalayas. The two species are similar in appearance.

35. *Pneopyga squamata.* The scaly-breasted wren. In shape and size this is very like the wren of England, but its upper plumage is not barred with black, as in the English species.

It is fairly common about Darjeeling, but is of retiring habits.

THE SYLVIIDÆ OR WARBLER FAMILY

36. *Abrornis superciliaris.* The yellow-bellied flycatcher-warbler.

A tiny bird about the size of a wren. The head is grey and the remainder of the upper plumage brownish yellow. The eyebrow is white, as are the chin, throat, and upper breast: the remainder of the lower plumage is bright yellow.

37. *Suya atrigularis.* The black-throated hill-warbler. The upper plumage is olive brown, darkest on the head. The chin, throat, breast, and upper abdomen are black.

THE LANIIDÆ OR SHRIKE FAMILY

38. *Lanius tephronotus.* The grey-backed shrike.

39. *Pericrocotus brevirostris.* The short-billed minivet. Very common about Darjeeling.

40. *Campophaga melanoschista.* The dark-grey cuckoo-shrike.

Plumage is dark grey, wings black, tail black tipped with white. Rather larger than a bulbul. Cuckoo-shrikes keep to trees, and rarely, if ever, descend to the ground.

THE MUSCICAPIDÆ OR FLYCATCHER FAMILY

Of the common flycatchers of the Western Himalayas, the following occur in the Eastern Himalayas:

41. *Stoparola melanops.* The verditer flycatcher. Very common at Darjeeling.

42. *Cyornis superciliaris.* The white-browed blue-flycatcher.

43. *Alseonax latirostris.* The brown flycatcher. Not very common.

44. *Niltava sundara.* The rufous-bellied niltava. Very abundant at Darjeeling. In addition to the rufous-bellied niltava, two other niltavas occur in the Eastern Himalayas.

45. *Niltava grandis.* The large niltava. This may be readily distinguished on account of its comparatively large size. It is as large as a bulbul. It is very common about Darjeeling.

46. *Niltava macgrigoriæ.* The small niltava. This is considerably smaller than a sparrow and does not occur above 5000 feet.

47. *Terpsiphone affinis.* The Burmese paradise flycatcher. This replaces the Indian species in the Eastern Himalayas, but it is not found so high up as Darjeeling, being confined to the lower ranges.

The other flycatchers commonly seen in the Eastern Himalayas are:

48. *Rhipidura allicollis.* The white-throated fantail flycatcher. This beautiful bird is abundant in the vicinity of Darjeeling. It is a black bird, with a white eyebrow, a whitish throat, and white tips to the outer tail feathers. It is easily recognised by its cheerful song and the way in which it pirouettes among the foliage and spreads its tail into a fan.

49. *Hemichelidon sibirica.* The sooty flycatcher. This is a tiny bird of dull brown hue which, as Jerdon says, has very much the aspect of a swallow.

50. *Hemichelidon ferruginea.* The ferruginous flycatcher. A rusty-brown bird (the rusty hue being most pronounced in the rump and tail) with a white throat.

51. *Cyornis rubeculoides.* The blue-throated flycatcher. The cock is a blue bird with a red breast. There is some black on the cheeks and in the wings.

The hen is a brown bird tinged with red on the breast. This species, which is smaller than a sparrow, keeps mainly to the lower branches of trees.

52. *Anthipes moniliger.* Hodgson's white-gorgeted flycatcher. A small reddish-brown bird with a white chin and throat surrounded by a black band,

that sits on a low branch and makes occasional sallies into the air after insects, can be none other than this flycatcher.

53. *Siphia strophiata*. The orange-gorgeted flycatcher. A small brown bird with an oval patch of bright chestnut on the throat, and some white at the base of the tail. (This white is very conspicuous when the bird is flying.) This flycatcher, which is very common about Darjeeling, often alights on the ground.

54. *Cyornis melanoleucus*. The little pied flycatcher. A very small bird. The upper plumage of the cock is black with a white eyebrow and some white in the wings and tail. The lower parts are white. The hen is an olive-brown bird with a distinct red tinge on the lower back. This flycatcher is not very common.

THE TURDIDÆ OR THRUSH FAMILY

55. *Oreicola ferrea*. The dark-grey bush-chat. Not so abundant in the Eastern as in the Western Himalayas.

56. *Henicurus maculatus*. The Western spotted forktail.

57. *Microcichla scouleri*. The little forktail. This is distinguishable from the foregoing by its very short tail. It does not occur commonly at elevations over 5000 feet.

58. *Rhyacornis fuliginosus*. The plumbeous redstart or water-robin. Not common above 5000 feet in the Eastern Himalayas.

59. *Merula boulboul*. The grey-winged ouzel.

60. *Petrophila cinclorhyncha*. The blue-headed rock-thrush.

61. *Oreocincla molissima*. The plain-backed mountain-thrush. This is the thrush most likely to be seen in the Eastern Himalayas. It is like the European thrush, except that the back is olive brown without any dark markings.

THE FRINGILLIDÆ OR FINCH FAMILY

62. *Hæmatospiza sipahi*. The scarlet finch. The cock is a scarlet bird, nearly as large as a bulbul, with black on the thighs and in the wings and tail.

The hen is dusky brown with a bright yellow rump. This species has a massive beak.

63. *Passer montanus.* The tree-sparrow. This is the only sparrow found at Darjeeling. It has the habits of the house-sparrow. The sexes are alike in appearance. The head is chestnut and the cheeks are white. There is a black patch under the eye, and the chin and throat are black. The remainder of the plumage is very like that of the house-sparrow.

THE HIRUNDINIDÆ OR SWALLOW FAMILY

64. *Hirundo rustica.* The common swallow.

65. *Hirundo nepalensis.* Hodgson's striated swallow.

THE MOTACILLIDÆ OR WAGTAIL FAMILY

66. *Oreocorys sylvanus.* The upland pipit. This is not very common east of Nepal.

THE NECTARINIDÆ OR SUNBIRD FAMILY

67. *Æthopyga nepalensis.* The Nepal yellow-backed sunbird. This replaces *Æthopyga scheriæ* in the Eastern Himalayas, and is distinguished by having the chin and upper throat metallic green instead of crimson. It is the common sunbird about Darjeeling.

THE DICÆIDÆ OR FLOWER-PECKER FAMILY

68. *Dicæum ignipectus.* The fire-breasted flower-pecker.

THE PICIDÆ OR WOODPECKER FAMILY

69. Of the woodpeckers mentioned as common in the Western Himalayas, the only one likely to be seen at Darjeeling is *Hypopicus hypererythrus*—the rufous-bellied pied woodpecker, and this is by no means common. The woodpeckers most often seen in the Eastern Himalayas are:

70. *Dendrocopus cathpharius.* The lesser pied woodpecker. A speckled black-and-white woodpecker about the size of a bulbul. The top of the head

and the sides of the neck are red in both sexes; the nape also is red in the cock.

71. *Gecinus occipitalis*. The black-naped green woodpecker. This bird, as its name implies, is green with a black nape. The head is red in the cock and black in the hen. This species is about the size of a crow.

72. *Gecinus chlorolophus*. The small Himalayan yellow-naped woodpecker. This species is distinguishable from the last by its small size, a crimson band on each side of the head, and the nape being golden yellow.

73. *Pyrrhopicus pyrrhotis*. The red-eared bay woodpecker. The head is brown. The rest of the upper plumage is cinnamon or chestnut-red with blackish cross-bars. There is a crimson patch behind each ear, which forms a semi-collar in the male. This species seeks its food largely on the ground.

In addition to the above, two tiny little woodpeckers much smaller than sparrows are common in the Eastern Himalayas. They feed on the ground largely. They are:

74. *Picumnus innominatus*. The speckled piculet.

75. *Sasia ochracea*. The rufous piculet. The former has an olive-green forehead. In the latter the cock has a golden-yellow forehead and the hen a reddish-brown forehead.

THE CAPITONIDÆ OR BARBET FAMILY

76. *Megalæma marshallorum*. The great Himalayan barbet.

77. *Cyanops franklini*. The golden-throated barbet. About the size of a bulbul. General hue grass green tinged with blue. The chin and throat are golden yellow. The forehead and a patch on the crown are crimson. The rest of the crown is golden yellow. The call has been syllabised as *kattak-kattak-kattak*.

THE ALCEDINIDÆ OR KINGFISHER FAMILY

78. *Ceryle lugubris*. The Himalayan pied kingfisher.

THE BUCEROTIDÆ OR HORNBILL FAMILY

Hornbills are to be numbered among the curiosities of nature. They are characterised by the disproportionately large beak. In some species this is nearly a foot in length. The beak has on the upper mandible an excrescence which in some species is nearly as large as the bill itself. The nesting habits are not less curious than the structure of hornbills. The eggs are laid in a cavity of a tree. The hen alone sits. When she has entered the hole she and the cock plaster up the orifice until it is only just large enough to allow the insertion of the hornbill's beak. The cock feeds the sitting hen during the whole period of her voluntary incarceration.

Several species of hornbills dwell in the forests at the foot of the Himalayas, but only one species is likely to be found at elevations above 5000 feet. This is the rufous-necked hornbill.

79. *Aceros nepalensis*. The rufous-necked hornbill. In this species the casque or excrescence on the upper mandible is very slight. It is a large bird 4 feet long, with a tail of 18 inches and a beak of 8½ inches. The hen is wholly black, save for a little white in the wings and tail. In the cock the head, neck, and lower parts are bright reddish brown. The rest of his plumage is black and white. In both sexes the bill is yellow with chestnut grooves. The naked skin round the eye is blue, and that of the throat is scarlet. The call of this species is a deep hoarse croak.

THE CYPSELIDÆ OR SWIFT FAMILY

80. *Cypselus affinis*. The common Indian swift.

81. *Chætura nudipes*. The white-necked spine-tail. A black bird glossed with green, having the chin, throat, and front and sides of the neck white.

THE CUCULIDÆ OR CUCKOO FAMILY

82. *Cuculus canorus*. The common or European cuckoo.

83. *Cuculus saturatus*. The Himalayan cuckoo.

84. *Cuculus poliocephalus*. The small cuckoo. This is very like the common cuckoo in appearance, but it is considerably smaller. Its loud unmusical call has been syllabised *pichu-giapo*.

85. *Cuculus micropterus*. The Indian cuckoo.

86. *Hierococcyx varius.* The common hawk-cuckoo.

87. *Hierococcyx sparverioides.* The large hawk-cuckoo.

THE PSITTACIDÆ OR PARROT FAMILY

88. *Palæornis schisticeps.* The slaty-headed paroquet. This bird is not nearly so common in the Eastern as in the Western Himalayas.

THE STRIGIDÆ OR OWL FAMILY

89. *Glaucidium brodei.* The collared pigmy owlet.

90. *Syrnium indrani.* The brown wood-owl.

91. *Scops spilocephalus.* The spotted Himalayan scops owl.

THE VULTURIDÆ OR VULTURE FAMILY

92. *Gyps himalayensis.* The Himalayan griffon.

93. *Pseudogyps bengalensis.* The white-backed vulture.

THE FALCONIDÆ OR FAMILY OF BIRDS OF PREY

94. *Aquila helica.* The imperial eagle.

95. *Hieraëtus fasciatus.* Bonelli's eagle.

96. *Ictinaëtus malayensis.* The black eagle. This is easily recognised by its dark, almost black, plumage.

97. *Spilornis cheela.* The crested serpent eagle.

98. *Milvus govinda.* The common pariah kite.

99. *Tinnunculus alaudaris.* The kestrel.

THE COLUMBIDÆ OR DOVE FAMILY

100. *Sphenocercus sphenurus.* The kokla green-pigeon.

101. *Turtur suratensis.* The spotted dove.

102. *Macropygia tusalia.* The bar-tailed cuckoo-dove.

THE PHASIANIDÆ OR PHEASANT FAMILY

103. *Gennæus leucomelanus.* The Nepal kalij pheasant. This is the only pheasant at all common about Darjeeling. It is distinguished from the white-crested kalij pheasant by the cock having a glossy blue-black crest. The hens of the two species resemble one another closely in appearance.

104. *Coturnix communis.* The grey quail.

105. *Arboricola torqueola.* The common hill partridge.

106. *Francolinus vulgaris.* The black partridge. Fairly common at elevations below 4000 feet.

THE CHARADRIIDÆ OR PLOVER FAMILY

107. *Scolopax rusticola.* The woodcock.

In the summer this bird is not likely to be seen below altitudes of 8000 feet above the sea-level.

TITS AT WORK

The average Himalayan house is such a ramshackle affair that it is a miracle how it holds together. The roof does not fit properly on to the walls, and in these latter there are cracks and chinks galore. Perhaps it is due to these defects that hill houses do not fall down more often than they do.

Thanks to their numerous cracks they do not offer half the resistance to a gale of wind that a well-built house would.

Be this as it may, the style of architecture that finds favour in the hills is quite a godsend to the birds, or rather to such of the feathered folk as nestle in holes. A house in the Himalayas is, from an avian point of view, a maze of nesting sites, a hotel in which unfurnished rooms are always available.

The sparrow usually monopolises these nesting sites. He is a regular dog-in-the-manger, for he keeps other birds out of the holes he himself cannot utilise. However, the sparrow is not quite ubiquitous. In most large hill stations there are more houses than he is able to monopolise.

I recently spent a couple of days in one of such, in a house situated some distance from the bazaar, a house surrounded by trees.

Two green-backed tits (*Parus monticola*) were busy preparing a nursery for their prospective offspring in one of the many holes presented by the building in question. This had once been a respectable bungalow, surrounded by a broad verandah. But the day came when it fell into the hands of a boarding-house keeper, and it shared the fate of all buildings to which this happens. The verandahs were enclosed and divided up by partitions, to form, in the words of the advertisement, "fine, large, airy rooms." There can be no doubt as to their airiness, but captious persons might dispute their title to the other epithets. A *kachcha* verandah had been thrown out with a galvanised iron roof and wooden supporting pillars. The subsequently-added roof did not fit properly on to that of the original verandah, and there was a considerable chink between the beam that supported it and the wall that enclosed the old verandah, so that the house afforded endless nesting sites. An inch-wide crack is quite large enough to admit of the passage of a tit; when this was negotiated the space between the old and the new roof afforded endless possibilities. Small wonder, then, that a pair of tits had elected to nest there.

The green-backed tit is one of the most abundant birds in the Himalayas. It is about the size of a sparrow. The head is black with a small perky crest. The cheeks are spotless white. The back of the head is connected by a narrow black collar with an expansive shirtfront of this hue. The remainder of the plumage is bright yellow. The back is greenish yellow, the rest of the plumage is slaty with some dashes of black and white. Thus the green-backed tit is a smart little bird. It is as vivacious as it is smart. It constantly utters a sharp, not unpleasant, metallic dissyllabic call, which sounds like *kiss me, kiss me, kiss me, kiss me*. This is one of the most familiar of the tunes that enliven our northern hill stations.

So much for the bird: now for its nest. A nest in a hole possesses many advantages. Its preparation does not entail very much labour. It has not to be built; it merely needs furnishing, and this does not occupy long if the occupiers have Spartan tastes. The tits in question were luxuriously inclined, if we may judge by the amount of moss that they carried into that hole. By the time it was finished it must have been considerably softer than the bed that was provided for my accommodation!

Moss in plenty was to be had for the taking; the trunks and larger branches of the trees which surrounded the "hotel" were covered with soft green moss. The tits experienced no difficulty in ripping this off with the beak.

The entrance to the nest hole faced downwards and was guarded on one side by the wall of the house, and on the other by a beam, so that it was not altogether easy of access even to a bird. Consequently a good deal of the moss gathered by the tits did not reach its destination; they let it fall while they were negotiating the entrance.

When a piece of moss dropped from the bird's beak, no attempt was made to retrieve it, although it only fell some 10 feet on to the floor of the verandah. In this respect all birds behave alike. They never attempt to reclaim that which they have let fall. A bird will spend the greater part of half an hour in wrenching a twig from a tree: yet, if this is dropped while being carried to the nest, the bird seems to lose all further interest in it.

By the end of the first day's work at the nest, the pair of tits had left quite a respectable collection of moss on the floor. This was swept away next morning. On the second day much less was dropped; practice had taught the tits how best to enter the nest hole.

It will be noticed that I speak of "tits." I believe I am correct in so doing; I think that both cock and hen work at the nest. I cannot say for certain, for I am not able to distinguish a lady- from a gentleman-tit. I never saw them together at the nest, but I noticed that the bird bringing material to it sometimes flew direct from a tree and at others alighted on the projecting end of a roof beam which the carpenters had been too lazy to saw off. It is my belief that the bird that used to alight on the beam was not the same as the one that flew direct from the tree. Birds are creatures of habit. If you observe a mother bird feeding her young, you will notice that she, when not disturbed, almost invariably approaches the nest in a certain fixed manner. She will perch, time after time, on one particular branch near the nest, and thence fly to her open-mouthed brood. When both parents bring food to the nest, each approaches in a way peculiar to itself; the hen will perhaps always come in from the left and the cock from the right.

The tits in question worked spasmodically at the nest throughout the hours of daylight. For ten minutes or so they would bring in piece after piece of moss at a great pace and then indulge in a little relaxation. All work and no play makes a tit a dull bird.

I had to leave the hotel late on the second day, so was not able to follow up the fortunes of the two little birds. I have, however, to thank them for affording me some amusement and giving me pleasant recollections of the place. It was good to lounge in a long chair, drink in the cool air, and watch the little birds at work. I shall soon forget the tumble-down appearance of the house, its seedy furniture, its coarse durries, and its hard beds, but shall long remember the great snow-capped peaks in the distance, the green moss-clad trees near about, the birds that sang in these, the sunbeams that played among the leaves, and, above all, the two little tits that worked so industriously at their nest.

THE PEKIN-ROBIN

This is not a robin, nor does it seem to be nearly related to the familiar redbreast; Pekin- or China-robin is merely the name the dealers give it, because a great many specimens are imported from China. Its classical name is *Liothrix lutea*. Oates calls it the red-billed liothrix. It is a bird about the size of a sparrow. The prevailing hue of the upper plumage is olive green, but the forehead is yellow. There is also a yellow ring round the eye, and the lower parts are of varying shades of this colour. Some of the wing feathers are edged with yellow and some with crimson, so that the wings, when closed, look as though lines of these colours are pencilled upon them. Oates, I notice, states that the hen has no red in the wing, but this does not seem to be the case in all examples. In the Pekin-robins that hail from China the chief difference between the sexes is that the plumage of the hen is a little duller than that of the cock. The bill is bright red. It is thus evident that the *liothrix* is a handsome bird, its beauty being of the quiet type which bears close inspection. But the very great charm of this sprightly little creature lies, not so much in its colouring, as in its form and movements. Its perfect proportions give it a very athletic air. In this respect it resembles the nimble wagtails. Next to these I like the appearance of the Pekin-robin better than that of any other little bird. Finn bestows even greater praise upon it, for he says: "Altogether it is the most generally attractive small bird I know of— everyone seems to admire it."

There is no bird more full of life. When kept in a cage, Pekin-robins hop from perch to perch with extraordinary agility, seeming scarcely to have touched one perch with their feet before they are off to another. I am inclined to think that the *liothrix*, like Camilla, Queen of the Volscians, could trip across a field of corn without causing the blades to move. This truly admirable bird is a songster of no mean capacity. Small wonder, then, that it has long been a favourite with fanciers. Moreover, it stands captivity remarkably well. It is the only insectivorous bird which is largely exported from India. So hardy is it that Finn attempted to introduce it into England, and with this object set free a number of specimens in St. James's Park some years ago, but they did not succeed in establishing themselves, although some individuals survived for several months. The English climate is to Asiatic birds much what that of the West Coast of Africa is to white men. J. K. Jerome once suggested that Life Insurance Companies should abolish the application form with its long list of queries concerning the ailments of the would-be insurer, his parents, grandparents, and other relatives, and substitute for it the German cigar test. If, said he, the applicant can come up smiling immediately after having smoked a German cigar, the Company could be certain that he was "a good life," to use the technical term. As regards birds, the survival of an English winter is an equally efficient test. The Pekin-robin is a very intelligent little bird. Finn found that it was not deceived by the resemblance between an edible and an unpalatable Indian swallow-tailed butterfly, although the sharp king-crow was deceived by the likeness.

Those Anglo-Indians who wish to make the acquaintance of the bird must either resort to some fancier's shop, or hie themselves to the cool heights of Mussoorie, or, better still, of Darjeeling, where the *liothrix* is exceptionally abundant. But even at Darjeeling the Pekin-robin will have to be looked for carefully, for it is of shy and retiring habits, and a small bird of such a disposition is apt to elude observation. In one respect the plains (let us give even the devil his due) are superior to the hills. The naturalist usually experiences little difficulty in observing birds in the sparsely-wooded flat country, but in the tree-covered mountains the feathered folk often require to be stalked. If you would see the Pekin-robin in a state of nature, go to some clearing in the Himalayan forest, where the cool breezes blow upon you direct from the snows, whence you can see the most beautiful sight in the world, that of snow-capped mountains standing forth against an azure sky. Tear your eyes away from the white peaks and direct them to the low bushes and trees which are springing up in the clearing, for in this you are likely to meet with a small flock of Pekin-robins. You will probably hear them before you see them. The sound to listen for is well described by Finn as "a peculiar five-noted call, *tee-tee-tee-tee-tee.*" As has been stated already, most, if not all, birds that go about in flocks in wooded country continually utter a call note, as it is by this means that the members of the flock keep together. Jerdon states that the food of the *liothrix* consists of "berries, fruit, seeds, and insects." He should, I think, have reversed the order of the bird's menu, for it comes of an insectivorous family—the babblers—and undoubtedly is very partial to insects—so much so that Finn suggests its introduction into St. Helena to keep them down. At the nesting season, in the early spring, the flock breaks up into pairs, which take upon themselves what Mr. E. D. Cuming calls "brow-wrinkling family responsibilities," and each pair builds in a low bush a cup-shaped nest.

BLACK BULBULS

All passerine birds which have hairs springing from the back of the head, and of which the tarsus—the lower half of the leg—is shorter than the middle toe, plus its claw, are classified by scientific men as members of the sub-family Brachypodinæ, or Bulbuls. This classification, although doubtless unassailable from the standpoint of the anatomist, has the effect of bringing together some creatures which can scarcely be described as "birds of a feather." The typical bulbul, as exemplified by the common species of the plains—Molpastes and Otocompsa—is a dear, meek, unsophisticated little bird, the kind of creature held up in copy-books as an example to youth, a veritable "Captain Desmond, V.C." Bulbuls of the nobler sort pair for life, and the harmony of their conjugal existence is rarely marred by quarrels; they behave after marriage as they did in the days of courtship: they love to sit on a leafy bough, close up against one another, and express their mutual admiration and affection by means of a cheery, if rather feeble, lay. They build a model nest in which prettily-coloured eggs are deposited. These they make but little attempt to conceal, for they are birds without guile. But, alas, their artlessness often results in a rascally lizard or squirrel eating the eggs for his breakfast. When their eggs are put to this base use, the bulbuls, to quote "Eha," are "sorry," but their grief is short-lived. Within a few hours of the tragedy they are twittering gaily to one another, and in a wonderfully short space of time a new clutch of eggs replaces the old one. If this shares the fate of the first set, some more are laid, so that eventually a family of bulbuls hatches out.

Such is, in brief, the character of the great majority of bulbuls; they present a fine example of rewarded virtue, for these amiable little birds are very abundant; they flourish like the green bay tree. As at least one pair is to be found in every Indian garden, they exemplify the truth of the saying, the meek "shall inherit the earth," and give a new meaning to the expression, "the survival of the fittest." There are, however, some bulbuls which are so unlike the birds described above that the latter might reasonably deny relationship to them as indignantly as some human beings decline to acknowledge apes and monkeys as poor relations. As we have seen, most bulbuls are inoffensive, respectable birds, that lead a quiet, domesticated life. The cock and hen are so wrapped up in one another as to pay little heed to the outer world. Not so the black bulbuls. These are the antithesis of everything bulbuline. They are aggressive, disreputable-looking creatures, who go about in disorderly, rowdy gangs. The song of most bulbuls consists of many pleasant, blithe tinkling notes; that of the black bulbul, or at any rate of the Himalayan black bulbul, is scarcely as musical as the bray of the ass. Most bulbuls are pretty birds and are most particular about their personal appearance. Black bulbuls are as untidy as it is possible for a bird to be. The two types of bulbul stand to one another in much the same relationship as does the honest Breton peasant to the inhabitant of the Quartier Latin in Paris.

Black bulbuls belong to the genus *Hypsipetes*. Three species occur in India—the Himalayan (*H. psaroides*), the Burmese (*H. concolor*), and the South Indian (*H. ganeesa*). All three species resemble one another closely in appearance. Take a king-crow (*Dicrurus ater*), dip his bill and legs in red ink, cut down his tail a little, dust him all over so as to make his glossy black plumage look grey and shabby, ruffle his feathers, apply a little *pomade hongroise* to the feathers on the back of his head, and make some of them stick out to look like a dilapidated crest, and you may flatter yourself that you have produced a very fair imitation of a black bulbul as it appears when flitting about from one tree summit to another. Closer inspection of the bird reveals the fact that "black" is scarcely the right adjective to apply to it. Dark grey is the prevailing hue of its plumage, with some black on the head and a quantity of brown on the wings and tail.

The Himalayan species has a black cheek stripe, which the other forms lack; but it is quite unnecessary to dilate upon these minute differences. I trust I have said sufficient to enable any man, woman, or suffragette to recognise a noisy black bulbul, and, as the distribution of each species is well defined and does not overlap that of the other species, the fact that a bird is found in any particular place at once settles the question of its species. The South Indian bird occurs only in Ceylon and the hills of South-west India; hence Jerdon called this species the Nilgiri or Ghaut black bulbul. Men of science in their wisdom have given the Himalayan bird the sibilant name of *Hypsipetes psaroides*. The inelegance of the appellation perhaps explains why the bird has been permitted to retain it for quite a long while unchanged.

I have been charged with unnecessarily making fun of ornithological nomenclature. As a matter of fact, I have dealt far too leniently with the peccadillos of the ornithological systematist. Recently a book was published in the United States entitled *The Birds of Illinois and Wisconsin*. Needless to state that while the author was writing the book, ornithological terminology underwent many changes; but the author was able to keep pace with these and with those that occurred while the various proofs were passing through the press. It was after this that his real troubles began. Several changes took place between the interval of the passing of the final proof and the appearance of the book, so that the unfortunate author in his desire to be up to date had to insert in each volume a slip to the effect that the American Ornithologists' Union had in the course of the past few days changed the name of no fewer than three genera; consequently the genus Glaux had again become Cryptoglaux, and the genera Trochilus and Coturniculus had become, respectively, Archilochus and Ammodramus! But we are wandering away from our black bulbuls. The hillmen call the Himalayan species the *Ban Bakra*, which means the jungle goat. Why it should be so named I have not an idea, unless it be because the bird habitually "plays the goat!"

Black bulbuls seem never to descend to the ground; they keep almost entirely to the tops of lofty trees and so occur only in well-wooded parts of the hills. When the rhododendrons are in flower, these birds partake very freely of the nectar enclosed within their crimson calyces. Now, I am fully persuaded that the nectar of flowers is an intoxicant to birds, and of course this will account, not only in part for the rowdiness of the black bulbuls, but for the pugnacity of those creatures, such as sunbirds, which habitually feed upon this stimulating diet. Black bulbuls, like sunbirds, get well dusted with pollen while diving into flowers after nectar, and so probably act the part of insects as regards the cross-fertilisation of large flowers. In respect of nesting habits, black bulbuls conform more closely to the ways of their tribe than they do in other matters. The nesting season is early spring. The nursery, which is built in a tree, not in a bush, is a small cup composed largely of moss, dried grass, and leaves, held together by being well smeared with cobweb. The eggs have a pink background, much spotted with reddish purple. They display a great lack of uniformity as regards both shape and colouring.

A WARBLER OF DISTINCTION

So great is the number of species of warbler which either visit India every winter or remain always in the country, so small and insignificant in appearance are these birds, so greatly do they resemble one another, and so similar are their habits, that even the expert ornithologist cannot identify the majority of them unless, having the skin in one hand and a key to the warblers in the other, he sets himself thinking strenuously. For these reasons I pay but little attention to the warbler clan. Usually when I meet one of them, I am content to set him down as a warbler and let him depart in peace. But I make a few exceptions in the case of those that I may perhaps call warblers of distinction—warblers that stand out from among their fellows on account of their architectural skill, their peculiar habits, or unusual colouring. The famous tailor-bird (*Orthotomus sartorius*) is the best known of the warblers distinguished on account of architectural skill. As a warbler of peculiar habits, I may cite the ashy wren-warbler (*Prinia socialis*), which, as it flits about among the bushes, makes a curious snapping noise, the cause of which has not yet been satisfactorily determined. As warblers of unusual colouring, the flycatcher-warblers are pre-eminent. In appearance these resemble tits or white-eyes rather than the typical quaker-like warblers.

Cryptolopha xanthoschista and Hodgson's grey-headed flycatcher-warbler are the names that ornithologists have given to a very small bird. But, diminutive though he be, he is heard, if not seen, more often than any other bird in all parts of the Western Himalayas. It is impossible for a human being to visit any station between Naini Tal and Murree without remarking this warbler. It is no exaggeration to state that the bird's voice is heard in every second tree. Oates writes of the flycatcher-warblers, "they are not known to have any song." This is true or the reverse, according to the interpretation placed on the word "song." If song denotes only sweet melodies such as those of the shama and the nightingale, then indeed flycatcher-warblers are not singers. Nevertheless they incessantly make a joyful noise. I can vouch for the fact that their lay is heard all day long from March to October. Before attempting to describe the familiar sound, I deem it prudent to recall to the mind of the reader the notice that once appeared in a third-rate music-hall:— "The audience are respectfully requested not to throw things at the pianist. He is doing his best." To say that this warbler emits incessantly four or five high-pitched, not very musical notes, is to give but a poor rendering of his vocal efforts, but it is, I fear, the best I can do for him. He is small, so that the volume of sound he emits is not great, but it is penetrating. Even as the cheery lay of the *Otocompsa* bulbuls forms the dominant note of the bird chorus in our southern hill stations, so does the less melodious but not less cheerful call of the flycatcher-warblers run as an undercurrent through the melody of the feathered choir of the Himalayas.

In what follows I shall speak of Hodgson's grey-headed flycatcher-warbler as our hero, because I shrink from constant repetition of his double double-barrelled name. I should prefer to give him Jerdon's name, the white-browed warbler, but for the fact that there are a score or more other warblers with white eyebrows. Our hero is considerably smaller than a sparrow, being only a fraction over four inches in length, and of this over one-third is composed of tail. The head and neck are grey, the former being set off by a cream-coloured eyebrow. Along the middle of the head runs a band of pale grey; this "mesial coronal band," as Oates calls it, is far more distinct in some specimens than in others. The remainder of the upper plumage is olive green, and the lower parts are bright yellow. Coloured plate, No. XX, in Hume and Henderson's *Lahore to Yarkand*, contains a very good reproduction of the bird. The upper picture on the plate represents our hero, the lower one depicting an allied species, Brook's grey-headed flycatcher-warbler (*C. Jerdoni*). It is necessary to state this because the book in question was written in 1873, since when, needless to say, the scientific names of most birds have undergone changes. The plate in question also demonstrates the slenderness of the foundation upon which specific differences among warblers rest.

Our hero is an exceedingly active little bird. He is ever on the move, and so rapid are his movements that to watch him for any length of time through field-glasses is no mean feat. He and his mate, with perhaps a few friends, hop about from leaf to leaf looking for quarry, large and small. The manner in which he stows away a caterpillar an inch long is a sight for the gods!

Sometimes two or three of these warblers attach themselves, temporarily at any rate, to one of those flocks, composed mainly of various species of tits and nuthatches, which form so well-marked a feature of all wooded hills in India. Hodgson's warblers are pugnacious little creatures. Squabbles are frequent. It is impossible to watch two or three of them for long without seeing what looks like one tiny animated golden fluff ball pursuing another from branch to branch and even from tree to tree.

The breeding season lasts from March to June. The nest is globular in shape, made of moss or coarse grass, and lined with some soft material, such as wool. The entrance is usually at one side. The nest is placed on a sloping bank at the foot of a bush, so that it is likely to escape observation unless one sees the bird flying to it. Three or four glossy white eggs are laid. Many years ago Colonel Marshall recorded the case of a nest at Naini Tal "at the side of a narrow glen with a northern aspect and about four feet above the pathway, close to a spring from which my *bhisti* daily draws water, the bird sitting fearlessly while passed and repassed by people going down the glen within a foot or two of the nest." At the same station I recently had a very different experience. Some weeks ago I noticed one of these warblers fly with a straw in its beak to a place on a steep bank under a small bush. I could not see what it was doing there, but in a few seconds it emerged with the bill empty. Shortly afterwards it returned with another straw. Having seen several pieces of building material carried to the spot, I descended the bank to try to find the nest. I could find nothing; the nest was evidently only just commenced. I then went back to the spot from which I had been watching the birds, but they did not return again. I had frightened them away. Individual birds of the same species sometimes differ considerably in their behaviour at the nesting season. Some will desert the nest on the slightest provocation, while others will cling to it in the most quixotic manner. It is never safe to dogmatise regarding the behaviour of birds. No sooner does an ornithologist lay down a law than some bird proceeds to break it.

THE SPOTTED FORKTAIL

"Striking" is, in my opinion, the correct adjective to apply to the spotted forktail (*Henicurus maculatus*). Like the paradise flycatcher, it is a bird which cannot fail to obtrude itself upon the most unobservant person, and, once seen, it is never likely to be forgotten. I well remember the first occasion on which I saw a spotted forktail; I was walking down a Himalayan path, alongside of which a brook was flowing, when suddenly from a rock in mid-stream there arose a black-and-white apparition, that flitted away, displaying a long tail fluttering behind it. The plumage of this magnificent bird has already been described.

As was stated above, this species is often called the hill-wagtail. The name is not a particularly good one, because wagtails proper occur in the Himalayas.

The forktail, however, has many of the habits of the true wagtail. I was on the point of calling it a glorified wagtail, but I refrain. Surely it is impossible to improve upon a wagtail.

In India forktails are confined to the Himalayas and the mountainous parts of Burma.

There are no fewer than eight Indian species, but I propose to confine myself to the spotted forktail. This is essentially a bird of mountain streams. It is never found far from water, but occurs at all altitudes up to the snow-line, so that, as Jerdon says, it is one of the characteristic adjuncts of Himalayan scenery. Indeed I know of few things more enjoyable than to sit, when the sun is shining, on the bank of a well-shaded burn, and, soothed by the soft melody of running water, watch the forktails moving nimbly over the boulders and stones with fairy tread, half-flight half-hop.

Forktails continually wag the tail, just as wagtails do, but not with quite the same vigour, possibly because there is so much more to wag!

Like wagtails, they do not object to their feet being wet, indeed they love to stand in running water.

Forktails often seek their quarry among the dead leaves that become collected in the various angles in the bed of the stream; when so doing they pick up each leaf, turn it over, and cast it aside just as the seven sisters do. They seem to like to work upstream when seeking for food. Jerdon states that he does not remember ever having seen a forktail perch; nevertheless the bird frequently flies on to a branch overhanging the brook, and rests there, slowly vibrating its forked tail as if in deep meditation.

Spotted forktails are often seen near the places where the *dhobis* wash clothes by banging them violently against rocks, hence the name dhobi-birds, by which they are called by many Europeans. The little forktail does not haunt the washerman's *ghat* for the sake of human companionship, for it is a bird that usually avoids man. The explanation is probably that the shallow pool in which the *dhobi* works and grunts is well adapted to the feeding habits of the forktail. I may here remark that in the Himalayas the washerman usually pursues his occupation in a pool in a mountain stream overhung with oaks and rhododendron trees, amid scenery that would annually attract thousands of visitors did it happen to be within a hundred miles of London. Not that the prosaic *dhobi* cares two straws for the scenery—nor, I fear, does the pretty little forktail. As I have already hinted, forktails are rather shy birds. If they think they are being watched they become restless and stand about on boulders, uttering a prolonged plaintive note, which is repeated at intervals of a few seconds. When startled they fly off, emitting a loud scream. But they are pugnacious to others of their kind, especially at the breeding season. I once saw a pair attack and drive away from the vicinity of their nest a Himalayan whistling-thrush (*Myiophoneus temmincki*)—another bird that frequents hill-streams, and a near relation of the Malabar whistling-thrush or idle schoolboy.

The nursery of the forktail, although quite a large cup-shaped structure, is not easy to discover; it blends well with its surroundings, and the birds certainly will not betray its presence if they know they are being watched. The nest is, to use Hume's words, "sometimes hidden in a rocky niche, sometimes on a bare ledge of rock overhung by drooping ferns and sometimes on a sloping bank, at the root of some old tree, in a very forest of club moss." I once spent several afternoons in discovering a forktail's nest which I was positive existed and contained young, because I had repeatedly seen the parents carrying grubs in the bill. My difficulty was that the stream to which the birds had attached themselves was in a deep ravine, the sides of which were so steep that no animal save a cat could have descended it without making a noise and being seen by the birds. Eventually I decorated my *topi* with bracken fronds, after the fashion of 'Arry at Burnham Beeches on the August bank holiday. Thus arrayed, I descended to the stream and hid myself in the hollow stump of a tree, near the place where I knew the nest must be. By crouching down and drawing some foliage about me, I was able to command a small stretch of the stream. My arrival was of course the signal for loud outcries on the part of the parent forktails. However, after I had been squatting about ten minutes in my *cache*, to the delight of hundreds of winged insects, the suspicions of the forktails subsided, and the birds began collecting food, working their way upstream. They came nearer and nearer, until one of them passed out of sight, although it was within 10 feet of me. It was thus evident that the nest was so situated that what remained of the tree-trunk obstructed my view of it. This was annoying, but I had one resource left, namely, to sit patiently until the sound of chirping told me that a parent bird was at the nest with food.

This sound was not long in coming, and the moment I heard it, up I jumped like a Jack-in-the-box, but without the squeak, in time to see a forktail leave a spot on the bank about 6 feet above the water. I was surprised, as I had the day before examined that place without discovering the nest. However, I went straight to the spot from which the forktail had flown, and found the nest after a little searching. The bank was steep and of uneven surface. Here and there a slab of stone projected from it and pointed downwards. Into a natural hollow under one of these projecting slabs a nest consisting of a large mass of green moss and liver-worts had been wedged. From the earth above the slab grew some ferns, which partially overhung the nest. Across the nest, a few inches in front of it, ran a moss-covered root. From out of the mossy walls of the nest there emerged a growing plant. All these things served to divert attention from the nest, bulky though this was, its outer walls being over 2 inches thick. The inner wall was thin—a mere lining to the earth. The nest contained four young birds, whose eyes were barely open. The young ones were covered with tiny parasites, which seemed quite ready for a change of diet, for immediately after picking up one of the young forktails, I found some thirty or forty of these parasites crawling over my hand!

There is luck in finding birds' nests, as in everything else. A few days after I had discovered the one above mentioned, I came upon another without looking for it. When I was walking along a hill-stream a forktail flew out from the bank close beside me, and a search of thirty seconds sufficed to reveal a well-concealed nest containing three eggs. These are much longer than they are broad. They are cream-coloured, mottled and speckled with tiny red markings.

THE NEST OF THE GREY-WINGED OUZEL

On several occasions this year (1910) I have listened with unalloyed pleasure to the sweet blackbird-like song of the grey-winged ouzel (*Merula boulboul*) at Naini Tal—a station in the Himalayas, consisting of over a hundred bungalows dotted on the well-wooded hillsides that tower 1200 feet above a mountain lake that is itself 6000 feet above the level of the sea. On the northern slope of one of the mountains on the north side of the Naini Tal lake, is a deep ravine, through which runs a little stream. The sides of the ravine are covered with trees—mainly rhododendron, oak, and holly.

On July 1st I went 1000 feet down this ravine to visit the nest of a spotted forktail (*Henicurus maculatus*) which I had discovered a week previously. Having duly inspected the blind, naked, newly-hatched forktails, I went farther down the stream to try to see something of a pair of red-billed blue magpies (*Urocissa occipitalis*).

The magpies were not at home that afternoon, and while waiting for them I caught sight of a bird among the foliage lower down the hill. At first I took this for a Himalayan whistling-thrush. I followed its movements through my field-glasses, and saw it alight on part of the gnarled and twisted trunk of a rhododendron tree. Closer inspection showed that the bird was a grey-winged ouzel. He had apparently caught sight of me, for his whole attitude was that of a suspicious bird with a nest in the vicinity. He remained motionless for several minutes.

As I watched him a ray of sunlight penetrated the thick foliage and fell upon the part of the tree where he was standing, and revealed to me that he was on the edge of a cunningly-placed nest.

The trunk of the rhododendron tree bifurcated about 20 feet above the ground; one limb grew nearly upright, the other almost horizontally for a few feet, and then broke up into five branches, or, rather, gave off four upwardly-directed branches, each as thick as a man's wrist, and then continued its horizontal direction, greatly diminished in size.

The four upwardly-directed branches took various directions, each being considerably twisted, and one actually curling round its neighbour. At the junction of the various branches lay the nest, resting on the flat surface, much as a large, shallow pill-box might rest in the half-closed palm of the hand of a man whose fingers were rugged and twisted with years of hard toil.

The upper part of the trunk was covered by a thick growth of green moss, and from it two or three ferns sprang.

As the exterior of the nest consisted entirely of green moss, it blended perfectly with its surroundings. From below it could not possibly have been seen. When I caught sight of it I was standing above it at the top of the ravine, and even then I should probably have missed seeing it, had not that ray of sunlight fallen on the nest and imparted a golden tint to the fawn-coloured plumage of the nestlings which almost completely filled the nest cup.

The situation of this nest may be said to be typical, although cases are on record of the nursery being placed on the ground at the root of a tree, or on the ledge of a rock. Many ouzels' nests are placed on the stumps of pollard trees, and in such cases the shoots which grow out of the stump often serve to hide the nest from view. The nests built by grey-winged ouzels vary considerably in structure. The commonest form is that of a massive cup, composed exteriorly of moss and lined with dry grass, a layer of mud being inserted between the moss and the grass lining. This mud layer does not invariably occur.

The cock ouzel remained for fully five minutes with one eye on me, and then flew off. I seized the opportunity to approach nearer the nest, and took up a position on the hillside level with it, at a distance of about 14 feet.

In a few minutes the hen bird appeared. Her prevailing hue is reddish brown, while the cock is black all over, save for some large patches of dark grey on the wings. In each sex the bill and legs are reddish yellow, the bill being the more brightly coloured. The hen caught sight of me and beat a hurried retreat, without approaching the nest.

The young ouzels kept very still; occasionally one of them would half raise its head. That was almost the only movement I noticed.

Presently the cock appeared, with his beak full of caterpillars. He alighted on a branch a few feet from the nest, where he caught sight of me; but instead of flying off as the hen had done, he held his ground and fixed his eye on me, no doubt swearing inwardly, but no audible sound escaped him.

Whenever I have watched a pair of birds feeding their young, I have almost invariably noticed that one of them is far more alarmed at my presence than the other. The ouzels proved no exception to the rule. In this case it was the cock who showed himself the bolder spirit. He remained watching me for fully ten minutes, his legs and body as immobile as those of a statue, but he occasionally turned his head to one side in order to obtain a better view of me; and I could then see, outlined against the sky, the wriggling forms of several caterpillars hanging from his bill. I hoped that he would pluck up courage to feed his youngsters before my eyes; but his heart failed him, for presently he flew to another tree a little farther away, whence he again contemplated me. After this he kept changing his position, never uttering a sound, and always retaining hold of the beakful of caterpillars. After a little the hen returned with her bill full of caterpillars, but she did not venture within 75 feet of the nest. I was not permitted to observe how long it would take the parental instinct to overcome the natural timidity of the birds. The sky suddenly became overcast, and a few minutes later I found myself enveloped in what the Scotch call a "wet mist." At certain seasons of the year rain storms come up as unexpectedly in the Himalayas as they do in the Grampians.

The rain put a final end to my observations on that nest, as I had to leave Naini Tal on the following day—an event which caused more sorrow to me than to the ouzels!

THE BLACK-AND-YELLOW GROSBEAK

The Indian grosbeaks are birds of limited distribution; they appear to be confined to the forests on the higher ranges of the Himalayas. Their most striking feature is the stout conical bill, which is an exaggeration of that of the typical finch, and is responsible for the bird's name. In one genus of grosbeak—*Mycerobas*—the bill is as deep as it is long, while in the other genus—*Pycnorhamphus*—it is nearly as massive. Three species belonging to this latter genus occur in India, namely, *P. icteroides*, the black-and-yellow grosbeak, found in the Western Himalayas; *P. affinis*, the allied grosbeak, found in Nepal, Sikkim, Tibet, and Western China; and *P. carneipes*, the white-winged grosbeak, which occurs all along the higher Himalayas.

There is only one Indian species of the other genus; this is known as the spotted-winged grosbeak (*Mycerobas melanoxanthus*), the localities in which this occurs are said to be "the Himalayas from the Hazara country to Sikkim at considerable elevations and Manipur."

The only Indian grosbeak which I have met in the flesh is the yellow-and-black species. This bird is common in the hills round about Murree, so that, when on ten days' leave there, I had some opportunity of studying its habits. It is a bird of the same size as the Indian oriole (*Oriolus kundoo*). The cock grosbeak, indeed, bears a striking resemblance to the black-headed oriole (*Oriolus melanocephalus*). His whole head, chin, throat, wings, shoulders, upper-tail-coverts, and thighs are black, the remainder of the plumage is a rich yellow, tinged with orange at the hind neck. Thus the colour and markings are almost identical with those of the black-headed oriole, the chief difference being that the latter has a little yellow in the wing. So great is the resemblance that the casual observer will, in nine cases out of ten, mistake the grosbeak for an oriole. The resemblance extends to size and shape, as the following table shows:

	Length of Bird.	Length of Tail.	Length of Wing.	Length of Tarsus.	Length of Beak.
Grosbeak	9.0 in.	3.7 in.	5.2 in.	1.0 in.	1.0 in.
Oriole	9.5 "	3.4 "	5.4 "	1.0 "	1.3 "

The hen grosbeak differs considerably in colour and marking both from the cock of her species and from the hen black-headed oriole. She is a dull ashy-grey bird, tinged faintly with yellowish red on the back and abdomen. Her wings and tail are black. The only young grosbeak that I have seen resembled the female in appearance, except that it had a yellow rump. It was being fed by a cock bird.

Grosbeaks live in forests, and go about either in couples or in small companies. They seem to feed largely on the ground, picking up insects. The beak of the finch tribe is adapted to a diet of seeds; nevertheless, many finches vary this food with insects. I saw a grosbeak seize, shake, and devour a caterpillar about two inches in length. Grosbeaks also eat berries and stone fruit. When disturbed they at once betake themselves to a tree, among the branches of which they are able to make their way with great agility. Grosbeaks are restless birds, always on the move, here to-day and gone to-morrow. The cock emits a call at frequent intervals. This is not easy to describe. It sounds something like *kiu kree*.

The nest is a cup-shaped structure, composed exteriorly of twigs, grass, and moss, and lined with stalks of maiden-hair fern and fine roots. It is usually placed high up in a fir tree. Colonel Rattray believes that the birds bring up two broods in the year. They lay first in May, and, as soon as the young are able to shift for themselves, a second nest is made. Thus in July both young birds at large and nests with eggs are likely to be seen. The eggs are not unlike those of the English hawfinch; the ground colour is pale greenish grey, blotched and spotted with blackish brown. Sometimes the markings occur chiefly at the broad end of the eggs.

The most striking feature of the black-and-yellow grosbeak, and that on which I wish particularly to dwell, is the extraordinary resemblance that the cock bird bears to the cock black-headed oriole. If this extended to the hen, and if the grosbeak were parasitic on the oriole, it would be held up as an example of mimicry. We should be told that owing to its resemblance to its dupe it was able to approach the nest without raising any suspicion and deposit its egg. But the grosbeak is not parasitic on the oriole, and it is the cock and not the hen that bears the resemblance; moreover, the black-headed oriole does not occur in the Himalayas, so that neither the grosbeak nor the oriole can possibly derive any benefit from this resemblance.

Now, cabinet zoologists are never tired of writing about mimicry. They assert that when organisms belonging to different families bear a close external resemblance, this resemblance has been brought about by natural selection. Having made this assertion, they expend reams of paper in demonstrating how one or both of the species benefits by the resemblance.

However, scientific books make no mention of the resemblance between the oriole and the grosbeak. The reason for this is, of course, that the resemblance in this instance cannot be a case of mimicry. Now, I regret to have to say that men of science take up the same attitude towards their theories as lawyers do regarding the cases they argue in Courts of Justice.

There would be no harm in taking up this attitude if men of science were to explain that they are acting the part of advocates, that they are fighting for a theory, and trying to persuade the world to accept this theory. It is because they masquerade as judges, and put forward a one-sided case as a matured judicial finding, that I take exception to their methods.

The trouble is that scientific men to-day form a brotherhood, a hierarchy, which lays claim to infallibility, or rather tacitly assumes infallibility.

They form a league into which none are admitted except those who take the oath of allegiance; and, of course, to expose the weakness of the scientific doctrines of the time is equivalent to violating the oath of allegiance. Now, the man of science who has to earn his living by his science, has either to join the league or run the risk of starving. This explains how a small coterie of men has things very much its own way; how it can lay down the law without fear of contradiction. If a man does arise and declines to accept the fiats of this league, it is not difficult for the members to combine and tell the general public that that man is a foolish crank, who does not know what he is talking about; and the public naturally accepts this dictum.

The only scientific men who, as a class, are characterised by humility are the meteorologists. I always feel sorry for the meteorologist. He has to predict the weather, and every man is able to test the value of these predictions. The zoologist, on the other hand, does not predict anything. He merely lays down the law to people who know nothing of law. He assures the world that he can explain all organic phenomena, and the world believes him.

As a matter of fact, zoology is quite as backward as meteorology. Those who do not wish to be deceived will do well to receive with caution all the zoological theories which at present hold the field. Before many years have passed all of them will have been modified beyond recognition. Most of them are already out of date.

There are doubtless good reasons for the colouring of both the grosbeak and the oriole; what these reasons are we know not. But as neither derives any benefit from the resemblance to the other, this *resemblance* cannot have been effected by natural selection. Now, if the unknown forces, which cause the various organisms to take their varied colours and forms, sometimes produce two organisms of different families which closely resemble one another, and the organisms in question are so distributed that neither can derive the slightest advantage in the struggle for existence from the resemblance, there is no reason why similar resemblances should not be produced in the case of organisms which occupy the same areas of the earth.

Thus it is quite possible that many so-called cases of mimicry are nothing of the kind.

The mere fact that one of the organisms in question may profit by the likeness is not sufficient to demonstrate that natural selection is responsible for the resemblance.

In this connection we must bear in mind that, according to the orthodox Darwinian theory, the resemblance must have come about gradually, and in its beginnings it cannot have profited the mimic *as a resemblance*.

So plastic are organisms, and so great is the number of living things in the earth, that it is not surprising that very similar forms should sometimes arise independently and in different parts of the globe. Several instances of this fortuitous resemblance are cited in Beddard's *Animal Colouration*; others are cited in *The Making of Species* by Finn, and myself.

Perhaps the most striking case is that of a cuckoo found in New Zealand, known as *Eudynamis taitensis*. This is a near relative of the Indian koel, which bears remarkable resemblance to an American hawk (*Accipiter cooperi*). Writing of this cuckoo, Sir Walter Buller says: "Not only has our cuckoo the general contour of Cooper's sparrow-hawk, but the tear-shaped markings on the underparts, and the arrow-head bars on the femoral plumes are exactly similar in both. The resemblance is carried still further, in the beautifully-banded tail and marginal wing coverts, and likewise in the distribution of colours and markings on the sides of the neck. On turning to Mr. Sharpe's description of the young male of this species in his catalogue of the Accipitres in the British Museum, it will be seen how many of the terms employed apply equally to our Eudynamis, even to the general words, 'deep brown above with a chocolate gloss, all the feathers of the upper surface broadly edged with rufous.' ... Beyond the general grouping of the colours there is nothing to remind us of our own Bush-hawk; and that there is no great protective resemblance is sufficiently manifested, from the fact that our cuckoo is persecuted on every possible occasion by the tits, which are timorous enough in the presence of a hawk."

These cases of chance resemblance should make us unwilling to talk about "mimicry," unless there is actual proof that one or other of the similar species benefits by the resemblance.

These cases, further, throw light on the origin of protective mimicry where it does exist.

Protective mimicry is usually said to have been brought about by the action of natural selection. This is not strictly accurate. Natural selection cannot cause two showy, dissimilar species to resemble one another; all it can do is to seize upon and perfect a resemblance that has been caused by the numerous factors that have co-operated to bring about all the diversity of organic life upon this earth.

THE GREAT HIMALAYAN BARBET

Barbets may be described as woodpeckers that are trying to become toucans. The most toucan-like of them all is the great Himalayan barbet (*Megalæma marshallorum*). Barbets are heavily-built birds of medium size, armed with formidable beaks, which they do not hesitate to use for aggressive purposes. As regards the nests they excavate, the eggs they lay, the pad that grows on the hocks of young birds, and their flight, they resemble their cousins the woodpeckers. But they are fruit-eating birds, and not insectivorous; it is this that constitutes the chief difference between them and the woodpeckers. Barbets are found throughout the tropical world. A number of species occur in India. The best known of these is the coppersmith, or crimson-breasted barbet (*Xantholæma hæmatocephala*), the little green fiend, gaudily painted about the head, which makes the hot weather in India seem worse than it really is by filling the welkin with the eternal monotone that resembles the sound of a hammer on a brazen vessel. Nearly as widely distributed are the various species of green barbet (*Thereiceryx*), whose call is scarcely less exasperating than that of the coppersmith, and may be described as the word *kutur* shouted many times and usually preceded by a harsh laugh or cackle.

The finest of all the barbets are the *Megalæmas*. The great Himalayan barbet attains a length of 13 inches. There is no lack of colour in its plumage. The head and neck are a rich violet blue. The upper back is brownish olive with pale green longitudinal streaks. The lower back and the tail are bright green. The wings are green washed with blue, brown, and yellow. The upper breast is brown, and the remainder of the lower plumage, with the exception of a scarlet patch of feathers under the tail, is yellow with a blue band running along the middle line. This bright red patch under the tail is not uncommon in the bird world, and, curiously enough, it occurs in birds in no way related to one another and having little or nothing in common as regards habits. It is seen in many bulbuls, robins, and woodpeckers, and in the pitta. The existence of these red under tail-coverts in such diverse species can, I think, be explained only on the hypothesis that there is an inherent tendency to variation in this direction in many species.

A striking feature of the great Himalayan barbet is its massive yellow bill, which is as large as that of some species of toucan. Although the bird displays a number of brilliant colours, it is not at all easy to distinguish from its leafy surroundings. It is one of those birds which are heard more often than seen.

Barbets are never so happy as when listening to their own voices. Most birds sing and make a joyful noise only at the nesting season. Not so the barbets; they call all the year round; even unfledged nestlings raise up the voices of infantile squeakiness.

The call of the great Himalayan barbet is very distinctive and easy to recognise, but is far from easy to portray in words. Jerdon described the call as a plaintive *pi-o, pi-o*. Hutton speaks of it as *hoo-hoo-hoo*. Scully syllabises it as *till-low, till-low, till-low*. Perhaps the best description of the note is that it is a mournful wailing, *pee-yu, pee-yu, pee-yu*. Some like the note, and consider it both striking and pleasant. Others would leave out the second adjective. Not a few regard the cry as the reverse of pleasant, and consider the bird a nuisance. As the bird is always on the move—its call at one moment ascends from the depths of a leafy valley and at the next emanates from a tree on the summit of some hill—the note does not get on one's nerves as that of the coppersmith does. Whether men like its note or not, they all agree that it is plaintive and wailing. This, too, is the opinion of hillmen, some of whom declare that the souls of men who have suffered injuries in the Law Courts, and who have in consequence died of broken hearts, transmigrate into the great Himalayan barbets, and that is why these birds wail unceasingly *un-nee-ow, un-nee-ow*, which means "injustice, injustice." Obviously, the hillmen have not a high opinion of our Law Courts!

Himalayan barbets go about in small flocks, the members of which call out in chorus. They keep to the top of high trees, where, as has been said, they are not easily distinguished from the foliage. When perched they have a curious habit of wagging the tail from side to side, as a dog does, but with a jerky, mechanical movement. Their flight is noisy and undulating, like that of a woodpecker. They are said to subsist exclusively on fruit. This is an assertion which I feel inclined to challenge. In the first place, the species remains in the Himalayas all the year round, and fruit must be very scarce there in winter. Moreover, Mr. S. M. Townsend records that a barbet kept by him in captivity on one occasion devoured with gusto a dead mouse that had been placed in its cage. Barbets nest in cavities in the trunks of trees, which they themselves excavate with their powerful beaks, after the manner of woodpeckers. The entrance to the nest cavity is a neat circular hole in a tree at heights varying from 15 to 50 feet. Most birds which rear their broods in holes enter and leave the nest cavity fearlessly, even when they know they are being watched by human beings, evidently feeling that their eggs or young birds are securely hidden away in the heart of the tree. Not so the *Megalæma*. It is as nervous about the site of its nest as a lapwing is. Nevertheless, on one occasion, when the nest of a pair of the great Himalayan barbets was opened out and found to contain an egg and a young bird, which latter was left unmolested, the parent birds continued to feed the young one, notwithstanding the fact that the nest had been so greatly damaged. The eggs are white, like those of all species which habitually nest in holes.

PART II

The Common Birds of the Nilgiris

THE COMMON BIRDS OF THE NILGIRIS

The avifauna of the Nilgiris is considerably smaller than that of the Himalayas. This phenomenon is easily explained. The Nilgiris occupy a far less extensive area; they display less diversity of climate and scenery; the lofty peaks, covered with eternal snow, which form the most conspicuous feature of the Himalayan landscape, are wanting in the Nilgiris.

The birds found in and about a Nilgiri hill station differ in character from those of the plains distant but a score of miles.

Of the common birds of the plains of Madras, the only ones that are really abundant on the Nilgiris are the black crow, the sparrow, the white-eye, the Madras bulbul, the myna, the purple sunbird, the tailor-bird, the ashy wren-warbler, the rufous-backed shrike, the white-browed fantail flycatcher, the Indian pipit, the Indian skylark, the common kingfisher, the pied crested cuckoo, the scavenger vulture, the Pondicherry vulture, the white-backed vulture, the shikra, the spotted dove, and the little brown dove.

The distribution of the avifauna of mountainous countries is largely a matter of elevation. At the base of the Nilgiris all the plains birds of the neighbourhood occur, and most of them extend some way up the hillsides. The majority, however, do not ascend as high as 1000 feet.

At elevations of 3000 feet the avifauna of the hills is already markedly different from that of the plains; nevertheless many of the hill species do not descend to this level, at any rate in the summer.

It is, therefore, necessary, when speaking of a plains bird as occurring or not occurring on the hills, to define precisely what is intended by this expression.

That which follows is written for people who visit the Nilgiri hill stations in the hot weather, and therefore the birds described are those which occur at elevations of 5500 feet and upwards in the summer. Those which visit the hills only in winter are either altogether ignored or given but the briefest mention.

This article does not deal exhaustively with the birds of the Nilgiris; it is merely a short account of the birds commonly seen in the higher regions of those hills during the summer months. To compile an exhaustive list would be easy. I refrain from doing so because a reader unacquainted with Indian ornithology would, if confronted by such a list, find it difficult to identify the common birds.

With this by way of introduction, I will proceed to describe the birds in question, dealing with them according to the classification adopted in the standard book on Indian ornithology—the bird volumes of the "Fauna of British India" series.

THE CORVIDÆ OR CROW FAMILY

This family is not nearly so well represented on the Nilgiris as it is in the Himalayas. The only crow found on the Nilgiris is the Indian corby (*Corvus macrorhynchus*)—the large black crow familiar to persons living in the plains. He, alas, is plentiful in the various hill stations; but it is some consolation that the grey-necked *Corvus* ceases from troubling those who seek the cool heights.

Like the grey-necked crow, the Indian tree-pie is not found at the Nilgiri hill stations—5000 feet appears to be the highest elevation to which he attains.

Of the tits only one species can be said to be common on the higher Nilgiris: this is the Indian grey tit (*Parus atriceps*)—a striking little bird, smaller than a sparrow. The head, throat, and neck are black, and a strip of this hue runs down the middle of the abdomen. The wings and tail are grey. The cheeks, the sides of the abdomen, and a patch on the back of the head are white. There is also a narrow white bar in the wing, and the grey tail is edged with white. The bird is found all over India, but is far more abundant on the hills than in the plains.

Another tit which, I believe, does not ascend so high as Ootacamund, but which is not uncommon in the vicinity of Coonoor is the southern yellow tit (*Machlolophus haplonotus*). This bird is not, as its name would seem to imply, clothed from head to foot in yellow. Its prevailing hues are green and brown. The head, breast, and upper abdomen are bright yellow, except the crown, crest, a broad streak behind the eye, and a band running from the chin to the abdomen, which are black. It is impossible to mistake this sprightly little bird, which is like the English tom-tit in shape. Tits are arboreal in habits; they seldom descend to the ground. Sometimes they go about in small flocks. They are supposed to live chiefly on insects, but most of them feed on fruit and seeds also, and the grey tit, alas, eats peas, among which it works sad havoc. The inhabitants of the Nilgiris call this last *Puttani kurivi*, which, I understand, means the pea-bird.

THE CRATEROPODIDÆ OR BABBLER FAMILY

This heterogeneous family is well represented in the Nilgiris.

The Madras seven sisters (*Crateropus griseus*) do not ascend the hills to any considerable height. But, of course there are seven sisters in the hills. Every part of India has its flocks of babblers. The Nilgiri babbler is a shy bird; it seems to dislike being watched. One might think it is aware that it is not so beautiful as it might be. But this cannot be the reason, because it has no objection to any person hearing its voice, which may be likened to the squeak of a rusty axle. This Nilgiri babbler does not enter gardens unless they are somewhat unkempt and contain plenty of thick bushes.

Mirabile dictu, this shy and retiring bird is none other than the jungle babbler (*Crateropus canorus*)—the common seven sisters or *sath bhai*—which in northern India is as bold and almost as confiding as the robin. No one has attempted to explain why the habits of this species on the Nilgiris should differ so much from those it displays in other places.

The southern scimitar-babbler (*Pomatorhinus horsfieldi*), like the jungle babbler on the Nilgiris, is a bird heard more often than seen.

Every person who has spent any time at Coonoor must be well acquainted with the notes of this species. A common call is a loud *ko-ko-ko-e-e-e*. Sometimes one bird calls *ko-ko-ko*, and another answers *ko-ee*. When the birds are feeding in company, they keep up a continual chatter, which is not unpleasing to the ear. When alarmed they give vent to a harsh cry of a kind characteristic of the babbler tribe. The scimitar-babbler is a bird nearly as big as a myna. It is of brownish hue and has a tail of moderate length. The breast and chin are pure white, and there is a white line running along each side of the head from front to back. The yellow beak is long and curved, hence the adjectival "scimitar." It is impossible to mistake the bird. The difficulty is to obtain anything more than a fleeting glimpse of it. It is so shy that it takes cover the instant it knows that it is being watched. It hops about in thick bushes with considerable address, much as a crow-pheasant does. It feeds on insects, which it picks off the ground or from leaves and trunks of trees. It uses the long bill as a probe, by means of which it secures insects lurking in the crevices of bark.

The Nilgiri laughing-thrush (*Trochalopterum cachinnans*) is a very common bird on the hills. Like the two species of babbler already described, it is a shy creature, living amid thick shrubs, from which it seldom ventures far. The head is slightly crested, the upper plumage, including the wings and tail, is olive brown. The head is set off by a white eyebrow. The under parts are chestnut. The beak and legs are black. Laughing-thrushes congregate in small flocks. They subsist chiefly on fruit. Their cry is loud and characteristic; it may be described as a bird's imitation of human laughter. Their cheerful calls are among the sounds heard most often at Ootacamund and Coonoor.

The Indian white-eye (*Zosterops palpebrosa*) is a bird that has puzzled systematists. Jerdon classed it among the tits, and its habits certainly justify the measure; but later ornithologists have not accepted the dictum "Manners makyth bird," and have placed the white-eye among the babblers.

The white-eye is a plump little bird, considerably smaller than a sparrow. The head and back are yellowish green, becoming almost golden in the sunlight. The wings and tail are brown. The chin, breast, and feathers under the tail are bright yellow, the abdomen is white. Round the eye is a ring of white feathers, interrupted in front by a black patch.

From this ring—its most striking feature—the bird has derived its name. The ring is very regular, and causes the bird to look as though it had been decorating its eye with Aspinall's best enamel.

White-eyes invariably go about in flocks; each member of the company utters unceasingly a cheeping note in order to keep his fellows apprized of his movements. These birds feed largely on insects, which they pick off leaves in truly tit-like manner, sometimes even hanging head downwards in order to secure a morsel.

The beautiful southern green-bulbul (*Chloropsis malabarica*) is numbered among the Crateropodidæ. It is not a true bulbul. It is common on the lower slopes of the Nilgiris, but does not often venture as high as Coonoor. A rich green bulbul-like bird with a golden forehead, a black chin and throat, and a patch of blue on the wing can be none other than this species.

The true bulbuls are also classified among the Crateropodidæ.

My experience is that the common bulbul of the plains—*Molpastes hæmorrhous*, or the Madras red-vented bulbul—is very rarely seen at the Nilgiri hill stations. Jerdon, likewise, states that it ascends the Nilgiris only up to about 6000 feet. Davison, however, declares that the bird begins to get common 4 miles from Ootacamund and is very numerous about Coonoor and all down the ghats. Be this as it may, the Madras red-vented bulbul is not the common bulbul of the Nilgiris. Its sweet notes are very largely, if not entirely, replaced by the yet sweeter and more cheery calls of the hill-bulbul. It will be labour lost to look up this name in Oates's ornithology, because it does not occur in that work. The smart, lively little bird, whose unceasing twittering melody gives our southern hill stations half their charm, has been saddled by men of science with the pompous appellation *Otocompsa fuscicaudata*. Even more objectionable is the English name for the pretty, perky bird. What shall I say of the good taste of those who call it the red-whiskered bulbul, as though it were a seedy Mohammedan who dips his grizzly beard in a pot of red dye by way of beautifying it? I prefer to call this bird the southern hill-bulbul. This name, I admit, leaves something to be desired, because the species is not confined to the hills. It is to be found in most places along the west coast. Nor is it the only bulbul living on the hills. The justification for the name is that if a census were taken of the bird-folk who dwell in our hill stations, it would show that *Otocompsa fuscicaudata* outnumbered all the crows, mynas, sparrows, flycatchers, and sunbirds put together. It is *the* bird of the southern hills. Every thicket, every tree—nay, every bush on the hills—has its pair of bulbuls. This species has distinctive plumage. Its most striking feature is a perky crest, which arises from the crown of the head and terminates in a forwardly-directed point, like Mr. Punch's cap. The crest is black and gives the bird a very saucy air. The wings and tail are dark brown, but each feather has a pale edge, which makes a pattern like scales on a fish. Below the eye is a brilliant patch of crimson. A similarly-coloured but larger patch is displayed at the base of the tail. The lower part of the cheek is white; this is divided off from the snowy breast by a narrow black band. The breast is, in its turn, separated from the greyish abdomen by a broad black band, which ornithologists term a collaret. Sometimes the collaret is interrupted in the middle. The hill-bulbul is a most vivacious bird. From dawn to sunset it is an example of perpetual motion. Its vocal cords are as active as its wings. The tinkling sounds of this bulbul form the dominant notes of the bird chorus. Husband and wife almost always move about in company. They flit from tree to tree, from bush to bush, plucking raspberries and other hill fruit as they pass. Bulbuls eat insects, but not when fruit is available. Like all birds bulbuls have large appetites. Recently I saw an Otocompsa devour three wild raspberries within as many minutes, each berry was swallowed at one gulp—a surprising feat, considering the small size of the bird's bill.

A bulbul's nest is a beautifully-shaped cup, usually placed in a bush at about 3 feet from the ground. As a rule, the bulbul selects an exposed site for its nest; in consequence many of the eggs are devoured by lizards. Crows in particular are addicted to young bulbuls, and take full advantage of the simplicity of the parent birds. Probably, three out of four broods never reach maturity. But the bulbul is a philosophic little bird. It never cries over broken eggs. If one clutch is destroyed it lays another.

The yellow-browed bulbul (*Iole icteria*) demands notice in passing, because it is common on the minor ranges. Its upper plumage is greenish yellow, the wings being darker than the back. The lower parts are canary yellow; the bird has also a yellow ring round the eye. Its note has been described as a soft, mellow whistle.

A very different bird is the southern or Nilgiri black bulbul (*Hypsipetes ganeesa*). This is an untidy-looking creature. Its crest is ragged. Its general hue is shabby black or brown, tinged with grey in places. The bill and feet are bright coral red. Black bulbuls utter a variety of notes, most of which are pleasing to the human ear, although they incline to harshness. The birds go about in flocks.

THE SITTIDÆ OR NUTHATCH FAMILY

Nuthatches are little climbing birds characterised by short tails. Like woodpeckers, they feed on insects, which they pick off the trunks and branches of trees. Unlike woodpeckers, however, they move about the trunks of trees with the head pointing indifferently downwards or upwards. The common nuthatch of the Nilgiris is the velvet-fronted blue nuthatch (*Sitta frontalis*). The upper plumage is dark blue, the cock having a velvety-black forehead and a black streak through the eye. The lower parts are creamy white. The bill is coral red. The note is a loud *tee-tee-tee*.

THE DICRURIDÆ OR DRONGO FAMILY

Several species of drongo or king-crow occur on the Nilgiris, but not one of them is sufficiently abundant to be numbered among the common birds of the hill stations.

THE SYLVIIDÆ OR WARBLER FAMILY

Of the warblers it may be said "their name is legion." So many species exist, and the various species are so difficult to differentiate, that the family drives most field ornithologists to the verge of despair. Many of the Indian warblers are only winter visitors to India. Eliminating these, only two warblers are entitled to a place among the common birds of the Nilgiris. These are the tailor-bird and the ashy wren-warbler.

At Coonoor the tailor-bird (*Orthotomus sartorius*) is nearly as abundant as it is in the plains. Oates, be it noted, states that this species does not ascend the hills higher than 4000 feet. As a matter of fact, the tailor-bird does not venture quite up to the plateau, but it is perfectly at home at all elevations below 6000 feet. This species may be likened to a wren that has grown a respectable tail. The forehead is ruddy brown, the back of the head is grey, the back is brown tinged with green. The lower plumage is a pale cream colour. There is a black patch or bar on each side of the neck, visible only when the bird stretches its neck to utter its loud *to-wee, to-wee, to-wee*. In the breeding season the shafts of the middle pair of tail feathers of the cock grow out beyond the rest. These projecting, bristle-like feathers render the cock easy of identification.

The ashy wren-warbler (*Prinia socialis*) is another "tiny brownie bird." The wings and tail are brown, the remainder of the upper plumage is the colour of ashes, the under parts are cream coloured. This warbler is a slight, loosely-built bird, and is easily distinguished from others of its kind by the curious snapping noise it makes as it flits from bush to bush. It occurs in pairs or singly. Davison remarks that it is "very fond of working its way up to some conspicuous post—to the top of one of the long flower-stalks of *Lobelia excelsa*, for instance—where it will halt for a minute or two, and then, after making a feeble attempt at a song, will dive suddenly in the brushwood and disappear."

THE LANIIDÆ OR SHRIKE FAMILY

Shrikes or butcher-birds are hawks in miniature, as regards habits if not in structure. With the exception of the brown shrike (*Lanius cristatus*), which is merely a winter visitor to India, the rufous-backed shrike (*L. erythronotus*) is the only butcher-bird common on the Nilgiris. The head of this species is pale grey, the back is of ruddy hue. The lower parts are white. The forehead and a broad band running through the eye are black. A bird having a broad black band through the eye is probably a shrike, and if the bird in question habitually sits on an exposed branch or other point of vantage, and from thence swoops on to the ground to secure some insect, the probability of its being a butcher-bird becomes a certainty.

Closely related to the shrikes are the minivets. Minivets are birds of tit-like habits which wander about in small flocks from place to place picking insects from the leaves of trees. They are essentially arboreal birds. I have never seen a minivet on the ground.

The common minivet of the Nilgiris is the orange minivet (*Pericrocotus flammeus*). The head and back of the cock are black. His wings are black and flame-colour, the red being so arranged as to form a band running lengthwise and not across the wing. The tail feathers are red, save the median pair, which are black. During flight the flashing red obliterates the black, so that the moving birds resemble tongues of flame and present a beautiful and striking spectacle. The hen is marked like the cock, but in her the red is replaced by bright yellow. This beautiful bird ceases to be abundant at elevations higher than Coonoor.

THE ORIOLIDÆ OR ORIOLE FAMILY

Both the Indian oriole (*Oriolus kundoo*) and the black-headed oriole (*O. melanocephalus*) occur on the Nilgiris, but on the higher ranges they are nowhere numerous. They therefore merit only passing notice.

THE STURNIDÆ OR STARLING FAMILY

The common myna of the Nilgiris is not *Acridotheres tristis* but *Æthiopsar fuscus*—the jungle myna. The casual observer usually fails to notice any difference between the two species, so closely do they resemble one another. Careful inspection, however, shows that the jungle myna has a little patch of feathers in front of the head over the beak. *Æthiopsar fuscus* has all the habits of the common myna. Like the latter, it struts about sedately in company with cattle in order to snatch up the grasshoppers disturbed by the moving quadrupeds. It feeds largely on the insects that infest the capsules of *Lobelia excelsa*, and is often to be seen clinging, like a tit, to the stem in order to secure the insects. Davidson gives these mynas a very bad character, he declares that they do immense damage to the fruit gardens on the Nilgiris, so that without the aid of nets, it is next to impossible to preserve pears from their depredations.

No other species of myna is common on the Nilgiris.

THE MUSCICAPIDÆ OR FLYCATCHER FAMILY

As in the Himalayas so on the Nilgiris the family of flycatchers is well represented. In one small Nilgiri wood I have come across no fewer than six species of flycatcher.

The beautiful little black-and-orange flycatcher (*Ochromela nigrirufa*) is a bird peculiar to the hills of Southern India.

The head and wings of the cock are black, the rest of the body is orange, of deeper hue on the back and breast than on the other parts. The portions of the plumage that are black in the cock are slaty brown in the hen. This flycatcher feeds on insects. But unlike most of its kind, it picks them off the ground more often than it secures them in the air.

It never takes a long flight, and almost invariably perches on a branch not more than two feet above the ground. It emits a low cheeping note—a *chur-r-r*, which is not unlike the sound made by some insects.

The Nilgiri blue-flycatcher (*Stoparola albicaudata*) is stoutly-built and a little larger than a sparrow. The male is clothed from head to tail in dark blue; his wife is more dingy, having a plentiful admixture of brownish grey in her plumage. Blue-flycatchers often occur in little flocks. They have the usual habits of their family, except that they seem sometimes to eat fruit.

A pretty little bird, of which the head, back, tail, and wings are deep blue, and the breast is orange fading into pale yellow towards the abdomen, is Tickell's blue-flycatcher (*Cyornis tickelli*). It has the characteristic habits of its tribe, and continually makes, from a perch, little sallies into the air after flying insects. But, more often than not it starts from one branch, and, having secured its quarry, alights on another. It sings a joyous lay, not unlike that of the fantail-flycatcher, but less sweet and powerful. It nests in a hole in a tree or bank, laying in May two or three eggs very thickly speckled with red spots.

The grey-headed flycatcher (*Culicicapa ceylonensis*) is a bird of somewhat sombre plumage. Its total length is only five inches, and of this half is composed of tail. The head is ashy grey, the back and wings are greenish; the lower plumage is bright yellow, but this is not conspicuous except when the bird is on the wing. This flycatcher has a loud song, which may be syllabised: *Think of me.... Never to be.*

The white-browed fantail-flycatcher (*Rhipidura albifrontata*), which delights the inhabitants of Madras with its cheerful whistle of five or six notes, occurs on the Nilgiris, but is there largely replaced by an allied species—the white-spotted fantail-flycatcher (*R. pectoralis*). The latter has all the habits of the former. Both make the same melody, and each has the habit of spreading out and erecting the tail whenever it settles on a perch after a flight. The white-spotted is distinguishable from the white-browed species by the white eyebrow being much narrower and less conspicuous. It is a black bird with a white abdomen, some white in the wings and tail, a few white spots on the chin, and the white eyebrow mentioned above.

The most beautiful of all the flycatchers is *Terpsiphone paradisi*—the paradise-flycatcher, or ribbon-bird, as it is often called. This is fairly abundant on the Nilgiris. The cock in the full glory of his adult plumage is a truly magnificent object. His crested head is metallic blue-black. This stands out in sharp contrast to the remainder of the plumage, which is as white as snow. Two of his tail feathers, being 12 inches longer than the others, hang down like satin streamers. Young cocks are chestnut instead of white. Birds in both phases of plumage breed. The hen has the metallic blue-black crested head, but she lacks the elongated tail feathers. Her plumage is chestnut, like that of the young cock. In both the hen and the young cock the breast is white. As "Eha" remarks, the hen looks very like a bulbul.

THE TURDIDÆ OR THRUSH FAMILY

This heterogeneous family includes thrushes, chats, robins, accentors, and dippers.

The southern pied bush-chat (*Pratincola atrata*) is one of the commonest and most familiar birds of the Nilgiris. It frequents gardens and is often found near houses: hence it is known as the hill-robin. The cock is clothed in black except the lower part of the back, the under parts, and a bar on the wing, which are white. Those parts that are black in the cock are brown in the hen, while her back and under parts are russet instead of white, but the white bar on the wing persists. This species lives on insects. It dwells in low shrubs and captures its quarry on the ground. It nests in a hole in a bank or well, lining the same with grass or hair. But summer visitors to the hills are not likely to come across the eggs, because these are usually hatched before May.

The Nilgiri blackbird (*Merula simillima*) is very like the blackbird of England. The plumage of the cock, however, is not so black, and the legs, instead of being brown, are reddish. Its charming song, with which all who have visited Ootacamund are familiar, is almost indistinguishable from that of its European cousin.

The Nilgiri thrush (*Oreocincla nilgirensis*) resembles the European thrush in appearance. Its upper plumage is pale brown, spotted with black and buff; its throat and abdomen are white with black drops. This bird has a fine powerful song, but he who wishes to hear it has usually to resort to one of the forests on the plateau of the Nilgiris.

THE PLOCEIDÆ OR WEAVER-BIRD FAMILY

This family includes the weaver-birds, famous for their wonderful hanging retort-shaped nests, and the munias, of which the amadavat or *lal* is familiar to every resident of India as a cage bird.

The weaver-birds do not ascend the hills, but several species of munia are found on the Nilgiris. Spotted munias (*Uroloncha punctulata*) are abundant in the vicinity of both Coonoor and Ootacamund. They occur in flocks on closely-cropped grassland. They feed on the ground. They are tiny birds, not much larger than white-eyes. The upper plumage is chocolate brown, becoming a rich chestnut about the head and neck, while the breast and abdomen are mottled black and white, hence the popular name. The black spots on the breast and abdomen cause these to look like the surface of a nutmeg grater; for that reason this munia is sometimes spoken of as the nutmeg-bird. The rufous-bellied munia (*Uroloncha pectoralis*) occurs abundantly a little below Coonoor, but does not appear to ascend so high as Ootacamund. Its upper parts are chocolate brown, save the feathers above the tail, which Oates describes as "glistening fulvous." The wings and tail are black, as are the cheeks, chin, and throat. The lower parts are pinkish brown. The stout bill is slaty blue. Like the spotted munia, this species is considerably smaller than a sparrow.

The Indian red-munia or red waxbill or *lal* (*Sporæginthus amandava*) is another very small bird. Its bill and eyes are bright red. Over its brown plumage are dotted many tiny white spots. There are also some large patches of red or crimson, notably one on the rump. The amount of crimson varies considerably; in the breeding season nearly the whole of the upper plumage of the cock is crimson. Amadavats go about in flocks and utter a cheeping note during flight. Their happy hunting grounds are tangles of long grass. Amadavats occur all over the Nilgiris.

THE FRINGILLIDÆ OR FINCH FAMILY

Finches are seed-eating birds characterised by a stout bill, which is used for husking grain.

The common sparrow (*Passer domesticus*) is the best known member of the finch family. Most of us see too much of him. He is to be observed in every garden on the Nilgiris, looking as though the particular garden in which he happens to be belongs to him. As a rule, sparrows nest about houses, but numbers of them breed in the steep cuttings on the road between Coonoor and Ootacamund.

The only other finch common on the Nilgiris is the rose-finch (*Carpodacus erythrinus*). This, however, is only a winter visitor: it departs from the Nilgiris in April and does not return until the summer season is over.

THE HIRUNDINIDÆ OR SWALLOW FAMILY

This family includes the swallows and the martins.

The swallows commonly found on the Nilgiris in summer are the Nilgiri house-swallow (*Hirundo javanica*) and the red-rumped or mosque swallow (*H. erythropygia*). I regret to have to state that Oates has saddled the latter with the name "Sykes's striated swallow"; he was apparently seduced by the sibilant alliteration!

Those two swallows are easily distinguished. The latter is the larger bird; its upper parts are glossy steel-blue, except the rump, which is of chestnut hue. The house-swallow has the rump glossy black, but it displays a good deal of red about the head and neck.

In the cold weather the European swallow and two species of martin visit the Nilgiris.

THE MOTACILLIDÆ OR WAGTAIL FAMILY

In the winter several kinds of wagtail visit the Nilgiris, but only one species remains all the year round. This is the beautiful pied wagtail (*Motacilla maderaspatensis*), of which the charming song must be familiar to all residents of Madras. On the Nilgiris the bird is not sufficiently common to require more than passing notice.

The pipits are members of the wagtail family. They have not the lively colouring of the wagtails, being clothed, like skylarks, in homely brown, spotted or streaked with dark brown or black. They have the wagtail trick of wagging the tail, but they perform the action in a half-hearted manner.

The two pipits most often seen on the Nilgiris in summer are the Nilgiri pipit (*Anthus nilgiriensis*) and the Indian pipit (*A. rufulus*). I know of no certain method of distinguishing these two species without catching them and examining the hind toe. This is much shorter in the former than in the latter species. The Nilgiri pipit goes about singly or in pairs, and, although it frequents grassy land, it usually keeps to cover and flies into a tree or bush when alarmed. It is confined to the highest parts of the Nilgiris. The Indian pipit affects open country and seems never to perch in trees.

THE ALAUDIDÆ OR LARK FAMILY

The Indian skylark (*Alauda gulgula*) is common on the Nilgiris. Wherever there is a grassy plain this species is found. Like the English skylark, it rises to a great height in the air, and there pours forth its fine song.

To the ordinary observer the Indian skylark is indistinguishable from its European congener.

The other common lark of the Nilgiris is the Malabar crested lark (*Galerita cristata*). This is in shape and colouring very like the Indian skylark, but is easily distinguished by the pointed crest that projects upwards and backwards from the hind part of the head. The crested lark has a pretty song, which is often poured forth when the bird is in the air. This species does not soar so high as the skylark. Like the latter, it frequents open spaces.

THE NECTARINIDÆ OR SUNBIRD FAMILY

A bird of the plains which is to be seen in every Nilgiri garden is the beautiful little purple sunbird (*Arachnecthra asiatica*). He flits about in the sunbeams, passing from flower to flower, extracting with his long tubular tongue the nectar hidden away in their calyces. He is especially addicted to gladioli. His head gets well dusted with yellow pollen, which he carries like a bee from one bloom to another. In the case of flowers with very deep calyces, he sometimes makes short cut to the honey by piercing with his sharp curved bill a hole in the side through which to insert the tongue. The cock purple sunbird needs no description. His glistening metallic plumage compels attention. He is usually accompanied by his spouse, who is earthy brown above and pale yellow below.

The other sunbird commonly seen in hill-gardens is one appropriately named the tiny sun bird or honeysucker (*Arachnecthra minima*), being less than two-thirds the size of a sparrow. As is usual with sunbirds, the cock is attired more gaily than the hen. He is a veritable feathered exquisite. Dame Nature has lavished on his diminutive body most of the hues to be found in her well-stocked paint-box. His forehead and crown are metallic green. His back is red, crimson on the shoulders. His lower plumage might be a model for the colouring of a Neapolitan ice-cream; from the chin downwards it displays the following order of colours: lilac, crimson, black, yellow. The hen is brown above, with a dull red rump, and yellow below.

The purple-rumped sunbird (*Arachnecthra zeylonica*), which is very abundant in and about Madras, does not ascend the Nilgiris above 3000 feet.

Loten's sunbird (*A. lotenia*) ventures some 2500 feet higher, and has been seen in the vicinity of Coonoor. This species is in colouring almost indistinguishable from the purple sunbird, but its long beak renders it unmistakable.

THE DICÆIDÆ OR FLOWER-PECKER FAMILY

Flower-peckers, like sunbirds, are feathered exquisites. The habits of the two families are very similar, save that flower-peckers dwell among the foliage of trees, while sunbirds, after the manner of butterflies, sip the nectar from flowers that grow near the ground.

Every hill-garden can boast of one or two flower-peckers. These are among the smallest birds in existence. They are as restless as they are diminutive. So restless are they that it is very difficult to follow their movements through field-glasses, and they are so tiny that without the aid of field-glasses it is difficult to see them among the foliage in which they live, move, and have their being. These elusive mites continually utter a sharp *chick-chick-chick.* Two species are common on the Nilgiris.

They are known as the Nilgiri flower-pecker (*Dicæum concolor*) and Tickell's flower-pecker (*D. erythrorhynchus*). The latter is the more numerous. Both are olive-green birds, paler below than above. Tickell's species has the bill yellow: in the other the beak is lavender blue.

THE PICIDÆ OR WOODPECKER FAMILY

Woodpeckers are birds that feed exclusively on insects, which they pick off the trunks of trees. They move about over the bark with great address. Whether progressing upwards, downwards, or sideways, the head is always pointed upwards.

For some reason or other there is a paucity of woodpeckers on the Nilgiris. The Indian Empire can boast of no fewer than fifty-four species; of these only six patronise the Nilgiris, and but two appear to ascend higher than 5000 feet. The only woodpecker that I have noticed in the vicinity of Coonoor is Tickell's golden-backed woodpecker (*Chrysocolaptes gutticristatus*). I apologise for the name; fortunately the bird never has to sign it in full. This woodpecker is a magnificent bird, over a foot in length, being 1½ inch longer than the golden-backed species found in Madras itself. The cock has a crimson crest, the sides of the head and neck and the under parts are white, relieved by black streaks that run longitudinally. The back and wings appear golden olive in the shade, and when the sun shines on them they become a beautiful coppery red. The lower part of the back is crimson. The tail is black. The hen differs from the cock in having the crest black. When these birds fly, their wings make much noise. The species utters a high-pitched but somewhat faint screaming note.

THE CAPITONIDÆ OR BARBET FAMILY

Barbets are tree-haunting birds characterised by massive bills. They have loud calls of two or three notes, which they repeat with much persistence. They nestle in trees, themselves excavating the nest cavity. The entrance to the nest is invariably marked by a neat round hole, a little larger than a rupee, in the trunk or a branch of a tree. The coppersmith is the most familiar member of the clan. It does not occur on the Nilgiris, but a near relative is to be numbered among the commonest birds of those hills, being found in every wood and in almost every garden. This bird is fully as vociferous as the coppersmith, but instead of crying, *tonk-tonk-tonk*, it suddenly bursts into a kind of hoarse laugh, and then settles down to a steady *kutur-kutur-kutur*, which resounds throughout the hillside. This call is perhaps the most familiar sound heard in the hills. This species is called the lesser green barbet (*Thereiceryx viridis*) to distinguish it from the larger green barbet of the plains (*T. zeylonicus*). It is a vivid green bird with a dull yellow patch, devoid of feathers, round the eye. There are some brown streaks on the breast.

THE ALCEDINIDÆ OR KINGFISHER FAMILY

The only kingfisher that occurs abundantly throughout the Nilgiris is the common kingfisher (*Alcedo ispida*). This bird is not much larger than a sparrow. The head and nape are blue with faint black cross-bars. The back is glistening pale blue and the tail blue of darker hue. The wings are greenish blue. The sides of the head are gaily tinted with red, blue, black, and white. The lower parts are rusty red. The bill is black and the feet coral red. The beautiful white-breasted kingfisher (*Halcyon smyrnensis*)—the large blue species with the chocolate-coloured head and white breast—occurs on the Nilgiris at all elevations, but is not nearly so abundant as its smaller relative.

THE CYPSELIDÆ OR SWIFT FAMILY

Four species of swift are to be seen on the Nilgiris; two of them are the fleetest birds in existence; these are the alpine swift (*Cypselus melba*) and the brown-necked spine-tail (*Chætura indica*). The former progresses with ease at the rate of 100 miles an hour: the latter can cover 125 miles, while the former is flying 100. If we poor human beings were possessed of the motive power of swifts we should think nothing of flying to England on ten days' casual leave. This may be possible a few years hence, thanks to the aeroplane; but even then the swifts will have the advantage as regards cheapness of transit. The lower parts of the alpine swift are white, while those of the spine-tail are rich brown. Hence the two species may be differentiated at a glance.

The edible-nest swiftlet (*Collocalia fuciphaga*) is the commonest swift on the Nilgiris. It is only about half the size of the species mentioned above, being less than 5 inches in length. In my opinion, this bird is misnamed the edible-nest swiftlet, because a considerable quantity of grass and feathers is worked into the nest, and I, for my part, find neither grass nor feathers edible. But *chacun à son gout*.

There is, however, an allied species—the little grey-rumped swiftlet (*C. francicia*)—found in the Andaman Islands—of which the nests are really good to eat. This species constructs its tiny saucer-shaped nursery entirely of its own saliva.

April and May are the months in which to seek for the nests of the Nilgiri swiftlet, and the insides of caves the places where a search should be made.

The fourth swift of the Nilgiris, the crested swift (*Macropteryx coronata*), is not sufficiently abundant to merit description in this essay.

THE CAPRIMULGIDÆ OR NIGHTJAR FAMILY

Nightjars, or goatsuckers, to give them their ancient and time-honoured name, are birds that lie up during the day in shady woods and issue forth at dusk on silent wing in order to hawk insects. The most characteristic feature of a nightjar is its enormous frog-like mouth; but it is not easy to make this out in the twilight or darkness, so that the observer has to rely on other features in order to recognise goatsuckers when he sees them on the wing, such as their long tail and wings, their curious silent fluttering flight, their dark plumage with white or buff in the wings and tail, their crepuscular and nocturnal habits, and their large size. Nightjars are as large as pigeons.

The common species of the Nilgiris is the jungle nightjar (*Caprimulgus indicus*). For a couple of hours after nightfall, and the same period before dawn in the spring, this bird utters its curious call—a rapidly-repeated *cuck-chug-chuck-chuck*.

Horsfield's nightjar (*C. macrurus*) is perhaps not sufficiently abundant on the Nilgiris to deserve mention in this essay. A bird which after dark makes a noise like that produced by striking a plank with a hammer can be none other than this species.

THE CUCULIDÆ OR CUCKOO FAMILY

The koel (*Eudynamis honorata*) occurs on the Nilgiris and has been shot at Ootacamund. It betrays its presence by its loud *ku-il, ku-il, ku-il*. The common cuckoo of the hills is the hawk-cuckoo (*Hierococcyx varius*) or brain-fever bird. Its crescendo *brain-fever*, BRAIN-FEVER, BRAIN-FEVER prevents any person from failing to notice it. It victimises laughing-thrushes and babblers. It has a large cousin (*H. sparverioides*), which also occurs on the Nilgiris, and which likewise screams *brain-fever* at the top of its voice. Both species are like sparrow-hawks in appearance. The handsome pied crested cuckoo (*Coccystes jacobinus*), which cuckolds the seven sisters, is a bird easy to identify. It has a conspicuous crest. The upper plumage is glossy black, save for a white wing bar and white tips to the tail feathers. The lower parts are white.

The common coucal or crow-pheasant (*Centropus sinensis*) is a cuckoo that builds a nest and incubates its eggs. It is as big as a pheasant, and is known as the Griff's pheasant because new arrivals in India sometimes shoot it as a game bird. If naturalists could show that this cuckoo derived any benefit from its resemblance to a pheasant, I doubt not that they would hold it up as an example of protective mimicry. It is a black bird with rich chestnut wings. The black tail is nearly a foot long. The coucal is fairly abundant on the Nilgiris.

THE PSITTACIDÆ OR PARROT FAMILY

The green parrots of the plains do not venture far up the slopes of the hills. The only species likely to be seen on the Nilgiris at elevations of 4000 feet and upwards is the blue-winged paroquet (*Palæornis columboides*). This is distinguishable from the green parrots of the plains by having the head, neck, breast, and upper back dove-coloured. It has none of the aggressive habits of its brethren of the plains. It keeps mainly to dense forests. Jerdon describes its cry as "mellow, subdued, and agreeable." It is the prima donna of the Psittaci.

Another member of the parrot family found on the Nilgiris is the Indian loriquet, or love-bird or pigmy parrot (*Loriculus vernalis*). This is a short-tailed bird about the size of a sparrow. It is grass green in colour, save for the red beak, a large crimson patch on the rump, and a small blue patch on the throat. This species does not obtrude itself on the observer. It is seen in cages more often than in a state of nature. It sleeps with the head hanging down after the manner of bats, hence Finn calls this pretty little bird the bat-parrot.

THE STRIGIDÆ OR OWL FAMILY

Owls, like woodpeckers, do not patronise the Nilgiris very largely. The only owl that commonly makes itself heard on those mountains is the brown wood-owl (*Syrnium indrani*). This is the bird which perches on the roof of the house at night and calls *to-whoo*.

Occasionally, especially round about Ootacamund, the grunting *ur-ur-ur-ur* of the brown fish-owl (*Ketupa zeylonensis*) disturbs the silence of the night on the Nilgiris.

THE VULTURIDÆ OR VULTURE FAMILY

Only four species of vulture occur on the hills of South India. One of these is the smaller white scavenger vulture (*Neophron ginginianus*), which is probably the ugliest bird in the world. Its plumage is dirty white, except the tips of the wings, which are black. The head is not bald, as is the case with most vultures; it is covered with projecting feathers that form an exceedingly bedraggled crest. The bill, the naked face, and the legs are yellow. This vulture is popularly known as the shawk or Pharaoh's chicken. Young scavenger vultures are sooty brown.

The other three vultures common on the Nilgiris are the Pondicherry vulture (*Otogyps calvus*), the long-billed vulture (*Gyps indicus*), and the white-backed vulture (*Pseudogyps bengalensis*). The first is easily identified by means of its white waistcoat, a patch of white on the thighs, and large red wattles that hang down like the ears of a blood-hound. With the above exceptions the plumage is black.

The long-billed vulture is of a uniform brown-grey colour.

The white-backed vulture is a dark brown, almost black, bird, with a white back and a broad white band on the under surface of each wing, which is very noticeable when the bird is soaring high in the air on the watch for carrion.

The two commonest vultures of the Nilgiris are the scavenger and the white-backed species.

THE FALCONIDÆ OR FAMILY OF BIRDS OF PREY

The raptores are not very strongly represented on the Nilgiris. The only two eagles likely to be seen are Bonelli's eagle (*Hieraëtus fasciatus*) and the black eagle (*Ictinaëtus malayensis*). The plumage of the latter is of much darker hue than that of the former.

Bonelli's eagle is a bold bird that works great havoc among tame pigeons. It sometimes carries off a barnyard fowl.

The black eagle is content with smaller quarry: young birds, rats, and snakes, seem to constitute the chief articles of its diet.

Needless to state, the common pariah kite (*Milvus govinda*) is found on the Nilgiris. This useful bird usually sails in graceful circles high overhead, looking for food. Its cry is not heard so frequently on those hills as in the Himalayas, the reason being the different configuration of the two ranges. The Nilgiris are undulating and downlike, hence the kites are able, while hovering higher than the summits of the hills, to see what is happening in the valleys. In the Himalayas they cannot do this, because the valleys are usually deep. The kites, therefore, sail there at a lower level than the hill-tops, and their plaintive *chee-hee-hee-hee-hee* is heard throughout the day. It is not a very cheerful sound, so that in this respect the Nilgiris have an advantage over the Himalayas.

The majority of the kites appear to migrate from the Nilgiris during the south-west monsoon.

The Brahmany kite (*Haliastur indus*)—the handsome kite with white head and breast and rich chestnut-red wings—is sometimes seen on the Nilgiris, but scarcely sufficiently often to merit a place among the common birds.

The three remaining raptores that are of frequent occurrence on the hills of South India are the shikra (*Astur badius*), the crested goshawk (*Lophospizias trivirgatus*), and the kestrel (*Tinnunculus alaudarius*). The shikra is very like the brain-fever bird in appearance. It is a little smaller than the common house-crow. The upper plumage is ashy grey. The tail is of the same hue, but with broad dark brown cross-bars. In young birds the breast is white with dark drops; in older birds the drops become replaced by wavy rust-coloured cross-bars. The eye is bright yellow, as is the cere or base of the beak. The crested goshawk may be described in brief as a large shikra with a crest.

The kestrel is the bird known in England as the windhover, on account of its habit of hovering in mid-air on rapidly-vibrating wings before pouncing on the lizard or other small fry, for which it is ever on the watch. This species is about the same size as the shikra. The head, neck, and tail are grey; the back and wings are dull red. The lower parts are cream-coloured, spotted with brown.

THE COLUMBIDÆ OR DOVE FAMILY

Jerdon's imperial pigeon (*Ducula cuprea*) is a beautiful bird 17 inches long, of which the tail accounts for 7 inches. The prevailing hue of this pigeon is grey. The head, breast, abdomen, and neck are suffused with lilac. The back and wings are olive brown. The legs are dull lake red, as is the bill, except the tip, which is blue. This fine bird is confined to dense forest; it is said to be fond of the wild nutmeg.

The Nilgiri wood-pigeon (*Alsocomus elphistonii*) is another forest-haunting bird. Its prevailing hue is dove grey, with a beautiful gloss on the back, which appears lilac in some lights and green in others. The only other ornament in its plumage is a black-and-white shepherd's plaid tippet. The wood-pigeon is as large as the imperial pigeon. Of the doves, that which is most often seen on the Nilgiris is the spotted dove (*Turtur suratensis*). This is easily distinguished from the other members of the family by its reddish wings spotted with dark brown and pale buff. The only other dove likely to be seen at the Nilgiri hill stations is the little brown dove (*T. cambayensis*), which utters a five-or-six-syllabled coo.

THE PHASIANIDÆ OR PHEASANT FAMILY

This important family includes the pea- and the jungle-fowl and the various pheasants.

The peacock is not found at altitudes above 4000 feet.

Jungle-fowl are abundant on the Nilgiris. He who keeps his eyes open may occasionally see one of these birds running across a road in the hills. This must not lead the observer to think that jungle-fowl spend most of their time in sprinting across roads. The fact of the matter is that the fowl tribe do not appreciate their food unless they have to scratch for it. Paths and roads are highly scratchable objects, hence they are largely resorted to for food; further, they are used for the purpose of the daily dust-bath in which every self-respecting fowl indulges. If these birds are disturbed when feeding or bathing, they do not make for the nearest cover as most other birds do: they insist on running across the road, thereby giving the grateful sportsman a clear shot. The domestic rooster has the same habit. So has the Indian child. To test the truth of these assertions, it is only necessary to drive briskly along a street at the side of which children or fowls are playing in perfect safety. At the sight of the horse, the child or hen, as the case may be, makes a dash for the far side of the road, and passes almost under the horse's nose. The fowl always gets across safely. The child is not so fortunate.

Two species of jungle-fowl have partitioned the Indian peninsula between them. The red species (*Gallus ferrugineus*) has appropriated the part of India which lies between Kashmir and the Godavery; while the grey jungle-fowl (*G. sonnerati*) has possessed itself of the territory south of the Godavery. The third jungle-fowl (*G. lafayetti*) has to be content with Ceylon, but the size of its name very nearly makes up for its deficiency in acres!

Davison is my authority for stating that the *Strobilanthes whitiani*, which constitutes the main undergrowth of many of the forests of the Nilgiris, seeds only once in about seven years, and that when this plant is seeding the grey jungle-fowl assemble in vast numbers to feed on the seed. They collect in the same way for the sake of bamboo seeds. The crow of the cock, which is heard chiefly in the morning and the evening, is not like that of the red jungle-fowl. It has been syllabised *kuk-kah-kah-kaha-kuk*. The call of the hen may be expressed by the syllables *kukkun-kukkun*.

The red spur-fowl (*Galloperdix spadicea*) is perhaps the most abundant game bird of the Nilgiris. It is quite partridge-like in shape. Both sexes have red legs and a patch of red skin round the eye. The feathers of the cock are dull red with blue edges, while those of the hen are black with broad buff margins. The cock may be described as a dull red bird with a grey head and some buff scale-like markings, and the hen as a grey bird, heavily barred with black.

The only quail commonly seen on the Nilgiris is the painted bush-quail (*Microperdix erythrorhynchus*). A bird in shape like a partridge, but not much larger than a sparrow, is probably this species. The prevailing hue is umber brown with coarse black blotches. The cock has the breast white and the head black with a white eyebrow. The head of the hen is dull red. The bill, legs, and feet of both sexes are red.

THE CHARADRIIDÆ OR PLOVER FAMILY

This very large family includes the plovers, sandpipers, and snipes. It is not very well represented on the Nilgiris. In winter snipe and woodcock visit those mountains and afford good sport to the human residents, but all have gone northward long before the summer visitors arrive.

Several species of sandpiper likewise visit the Nilgiris in winter; one of these—the wood sandpiper (*Totanus glareola*)—tarries on until after the beginning of summer. This is a bird as large as a dove; its plumage is speckled brown and white. It looks somewhat like a snipe with a short bill. It lives on the margins of ponds and constantly wags its apology for a tail.

THE RALLIDÆ OR RAIL FAMILY

The rails are not well represented on the Nilgiris.

The water-hen (*Gallinula chloropus*) is common on the lake at Ootacamund. This is an olive-green bird about the size of a pigeon. Its bill and forehead are red; there is a patch of white under the tail. This species swims like a duck.

Another rail which may be seen sometimes in the Botanical Gardens at Ootacamund is the white-breasted water-hen (*Amaurornis phoenicurus*). This is a black bird with the face, throat, and breast white. There is a chestnut-hued patch under the tail.

THE ARDEIDÆ OR HERON FAMILY

Almost the only member of the heron family that visits the Nilgiri hill stations is the pond-heron or paddy-bird (*Ardeola grayii*).

A colony of these birds pursues its avocations on the margin of the lake at Ootacamund, but I believe that I am right in saying that the paddy-birds of Ootacamund go to the plains for nesting purposes.

PART III

The Common Birds of the Palni Hills

THE COMMON BIRDS OF THE PALNI HILLS

For the benefit of those who visit Kodikanal I have compiled a list of the birds most commonly seen at altitudes of over 5000 feet in the Palni hills. I must here state that I have no first-hand knowledge of the avifauna of those hills, and the list that follows is based on the observations of Dr. Fairbank, made nearly 40 years ago.

The avifauna of the Palni is a comparatively restricted one: which is in part doubtless explained by the comparatively small area of the higher ranges that is covered by forest.

The great majority of the birds that follow have been described in the chapter on the birds of the Nilgiris, and I have contented myself with merely naming such.

THE CORVIDÆ OR CROW FAMILY

1. *Corvus macrorhynchus*. The Indian corby. This is not very abundant above 5500 feet.

2. *Dendrocitta rufa*. The tree-pie. This does not appear to occur above 5000 feet.

3. *Machlolophus haplonotus*. The southern yellow tit. Occurs at Kodikanal, but is not very common there.

THE CRATEROPODIDÆ OR BABBLER FAMILY

4. *Crateropus canorus*. The jungle babbler. This rarely ascends higher than 5000 feet.

5. *Trochalopterum fairbanki*. The Palni laughing-thrush. This species is peculiar to the Palnis and the Anamallis. The head is very dark brown, almost black, with a broad white eyebrow. The cheeks are grey, as are the chin, throat, and breast. The back, wings, and tail are olive brown tinged with rusty red. The abdomen is bright rufous. The noisy cries of this bird are among the most familiar sounds of Kodikanal. It is destructive to peaches and raspberries.

6. *Pomatorhinus horsfieldi*. The southern scimitar-babbler. This is not nearly so abundant on the Palnis as on the Nilgiris.

7. *Zosterops palpebrosa.* The Indian white-eye. A common bird.

8. *Iole icteria.* The yellow-browed bulbul. *Otocompsa fuscicaudata.* The southern red-whiskered bulbul or hill-bulbul. As in the Nilgiris so in the Palnis, this is the most abundant bird on the higher hills.

9. *Molpastes hæmorrhous.* The Madras red-vented bulbul. The higher one ascends, the rarer this bird becomes.

10. *Hypsipetes ganeesa.* The southern black bulbul.

11. *Myiophoneus horsfieldi.* The Malabar whistling-thrush or idle schoolboy. This fine but shy bird is found on the streams up to 6000 feet. It is a bird as large as a crow, with glossy black plumage, in which are patches of bright cobalt blue.

It is better known to the ear than to the eye. It emits a number of cheerful whistling notes.

THE SITTIDÆ OR NUTHATCH FAMILY

12. *Sitta frontalis.* The velvet-fronted blue nuthatch. This bird is found in every part of the Palnis where there are trees.

THE DICRURIDÆ OR DRONGO FAMILY

13. *Chaptia ænea.* The bronzed drongo. This species is not often seen at altitudes of more than 5000 feet above sea-level.

It is like the common king-crow in appearance, but the plumage is glossed with a bronze sheen, and the tail is less markedly forked.

THE SYLVIIDÆ OR WARBLER FAMILY

14. *Orthotomus sartorius.* The tailor bird. This has been seen as high as 5500 feet above the sea-level.

15. *Prinia socialis.* The ashy wren-warbler.

16. *Prinia inorata.* The Indian wren-warbler. This is very like the ashy wren-warbler in appearance. Its upper plumage is earthy-brown, and not reddish brown, and it does not make during flight the curious snapping noise so characteristic of *P. socialis.*

THE LANIIDÆ OR SHRIKE FAMILY

17. *Lanius erythronotus*. The rufous-backed shrike.

18. *Pericrocotus flammeus*. The orange minivet. This beautiful bird occurs from the bottom to the top of the Palnis.

19. *Pericrocotus peregrinus*. The little minivet. This is a bird of the plains rather than of the hills. But as Fairbank observed it in the Palnis as high as 5000 feet, it is given a place in this list. *Cock*: Head and shoulders slaty grey, lower back deep scarlet, wings black with red bar, tail black with red at tip, chin and throat blackish, breast scarlet; lower plumage orange yellow. *Hen*: upper parts grey, lower parts creamy white, wing brown with yellow or orange bar, tail black with red tip.

This species is smaller than a sparrow, but the tail is 3 inches long.

THE ORIOLIDÆ OR ORIOLE FAMILY

20. *Oriolus melanocephalus*. The black-headed oriole. This species has been seen as high as 5000 feet above the sea-level. The cock is bright yellow, with a black head and some black in the wings and tail. The hen is of a much duller yellow and has the back tinged with green.

THE STURNIDÆ OR STARLING FAMILY

Fairbank does not mention the jungle myna (*Æthiopsar fuscus*) in his list of the birds of the Palnis (*Stray Feathers*, vol. v, 1877). Yet this is precisely the myna one would expect to find on the Palnis, and it should be looked for.

21. On the other hand, the Brahmany myna (*Temenuchus pagodarum*), which is essentially a bird of the plains, is said by Fairbank to occur "well up the hillsides."

Of the common myna (*Acridotheres tristis*), he writes: "This is common around villages at 4000 feet."

22. *Temenuchus pagodarum*. The Brahmany myna. Head and recumbent crest black. Wings black and grey. Tail brown with a white tip. Remainder of plumage rich buff. Beak blue with yellow tip. Legs bright yellow.

THE EULABETIDÆ OR GRACKLE FAMILY

23. *Eulabes religiosa.* The southern grackle or hill-myna. This bird occurs in the forests of the Palnis between elevations of 4000 and 5000 feet. It is familiar to every one as a cage bird. A glossy black bird with a white wing bar. The wattles, legs, and bill are yellow.

THE MUSCICAPIDÆ OR FLYCATCHER FAMILY

24. *Ochromela nigrirufa.* The black-and-orange flycatcher.

25. *Stoparola albicaudata.* The Nilgiri blue-flycatcher.

26. *Cyornis tickelli.* Tickell's blue-flycatcher. Less common than on the Nilgiris.

27. *Culicicapa ceylonensis.* The grey-headed flycatcher.

28. *Rhipidura albifrontata.* The white-browed fantail flycatcher. Fairbank did not find this bird at altitudes over 4000 feet.

THE TURDIDÆ OR THRUSH FAMILY

29. *Pratincola atrata.* The southern pied bush-chat or hill-robin. Not nearly so abundant on the Palnis as on the Nilgiris.

30. *Merula simillima.* The Nilgiri blackbird. In spring its delightful song gladdens the groves of the higher Palnis.

31. *Copschychus saularis.* The magpie-robin. Has been observed as high as 5000 feet. The cock is black, and the hen grey, with a white breast and white in the wings and tail. The distribution of the black and white is like that in the common magpie.

THE FRINGILLIDÆ OR FINCH FAMILY

32. *Passer domesticus.* The common sparrow. Does not occur much above 5000 feet.

THE HIRUNDINIDÆ OR SWALLOW FAMILY

33. *Hirunda javanica.* The Nilgiri house-swallow.

THE MOTACILLIDÆ OR WAGTAIL FAMILY

34. *Anthus nilgirensis.* The Nilgiri pipit. Common on the grassy fields at the summit of the Palnis.

THE NECTARINIDÆ OR SUNBIRD FAMILY

35. *Arachnecthra minima.* The tiny sunbird or honeysucker. Common from 4000 feet upwards.

THE DICÆIDÆ OR FLOWER-PECKER FAMILY

36. *Dicæum concolor.* The Nilgiri flower-pecker. This frequents the flowers of the parasitic *Loranthus.*

37. *Dicæum erythrorhynchus.* Tickell's flower-pecker. This species does not appear to ascend the Palnis to any great height. It is abundant at the foot of the hills.

THE PICIDÆ OR WOODPECKER FAMILY

38. *Chrysocolaptes gutticristatus.* Tickell's golden-backed woodpecker. As in the Nilgiris so in the Palnis, this is the common woodpecker.

39. *Brachypternus aurantius.* The golden-backed woodpecker. This is the common woodpecker of the plains: it ascends the Palnis to elevations of 5000 feet. This is distinguishable from the foregoing species by its smaller size, and in having the rump velvety black instead of crimson.

40. *Liopicus mahrattensis.* The yellow-fronted pied woodpecker. This plains species ascends the Palnis to elevations of 5000 feet. It is much smaller than either of the two foregoing species. The plumage is spotted black and white, with a patch of red on the abdomen. There is a yellow patch on the forehead. The cock has a short red crest.

THE CAPITONIDÆ OR BARBET FAMILY

41. *Thereiceryx viridis*. The small green barbet. (The coppersmith does not ascend higher than 4000 feet.)

THE ALCEDINIDÆ OR KINGFISHER FAMILY

42. The only kingfisher found in the Palnis seems to be the white-breasted kingfisher (*Halcyon smyrnensis*), but this species is confined to the lower hills.

THE UPUPIDÆ OR HOOPOE FAMILY

43. The Indian hoopoe (*Upupa indica*) occurs on the lower ranges, but does not appear to ascend the hills as far as Kodikanal.

THE CYPSELIDÆ OR SWIFT FAMILY

44. Swifts are not abundant in the Palnis. The only one observed by Fairbank was the common Indian swift (*Cypselus affinis*), seen at an elevation of 3000 feet. This is easily distinguished by the white band across the rump.

THE CUCULIDÆ OR CUCKOO FAMILY

45. *Hierococcyx varius*. The hawk-cuckoo.

46. *Eudynamis honorata*. The Indian koel. This species is not common on the Palnis.

47. *Centropus sinensis*. The common coucal or crow-pheasant. This is not very common.

THE PSITTACIDÆ OR PARROT FAMILY

48. *Palæornis columboides*. The blue-winged paroquet.

49. *Loriculus vernalis*. The Indian loriquet or love-bird.

THE STRIGIDÆ OR OWL FAMILY

50. *Ketupa zeylonensis.* The brown fish-owl. A large bird with aigrettes. The eyes are bright yellow. The legs are devoid of feathers. The call is a series of grunts.

THE VULTURIDÆ OR VULTURE FAMILY

51. *Neophron ginginianus.* The smaller white scavenger vulture. This occurs up to at least 5000 feet. Fairbank did not observe any other vultures on the higher hills, but it is unlikely that *Pseudogyps bengalensis* (the white-backed vulture), *Gyps indicus* (the long-billed vulture), and *Otogyps calvus* (the black or Pondicherry vulture) do not visit the higher hills. These three birds should be looked for, especially the first.

THE FALCONIDÆ OR FAMILY OF BIRDS OF PREY

52. *Ictinaëtus malayensis.* The black eagle. Not very common.

53. *Milvus govinda.* The common pariah kite. Fairbank did not see this above 3000 feet.

54. *Haliastur indus.* The Brahmany kite. Occurs up to at least 4000 feet.

55. *Tinnunculus alaudarius.* The kestrel.

THE COLUMBIDÆ OR DOVE FAMILY

56. *Alsocomus elphistonii.* The Nilgiri wood-pigeon.

The spotted and the little brown doves (*Turtur suratensis* and *T. cambayensis*) are found only on the lower hills.

THE PHASIANIDÆ OR PHEASANT FAMILY

57. *Gallus sonnerati.* The grey jungle fowl. Not so common as on the Nilgiris.

58. *Galloperdix spadicea.* The red spur-fowl. Not common.

59. *Microperdix erythrorhynchus*. The painted bush-quail.

THE CHARADRIIDÆ OR PLOVER FAMILY

A few snipe and woodcock visit the Palnis in winter.

THE PODICIPEDIDÆ OR GREBE FAMILY

60. *Podicipes albipennis*. The little grebe or dabchick. This bird never leaves the water. It is smaller than a dove. It has no tail. It is dark glossy brown in colour with chestnut on the sides of the neck.

APPENDICES

I. Vernacular Names of Himalayan Birds

II. Vernacular Names of Nilgiri Birds

I. VERNACULAR NAMES OF HIMALAYAN BIRDS

Ababil	swallow
Akku	common cuckoo
Argul	lammergeyer
Ban-bakra	black bulbul, rusty-cheeked scimitar-babbler
Ban-sarrah	black-throated jay
Ban-titar	hill partridge
Bara bharao	large hawk-cuckoo
Batasi	Indian swift
Bater	quail
Bhimraj	racquet-tailed drongo
Boukotako	Indian cuckoo
Bulaka	brown wood-owl
Bulbul	bulbul
Bunchil	cheer pheasant
Chakru	chakor partridge
Chaman	cheer pheasant
Chanjarol	woodcock
Chil	kite
Chir	cheer pheasant

Chitla	spotted dove
Chitroka fakhta	spotted dove
Chota fakhta	little brown dove
Chukar	chakor partridge
Digg-dall	blue magpie
Dhal kowa	corby
Dhor fakhta	ring-dove
Dogra chil	crested serpent eagle
Durkal	black bulbul
Gagi	slaty-headed paroquet
Gidh	vulture
Gir-chaondia	white-capped redstart
Gonriya	house-sparrow
Gugi	ring-dove
Herril	cheer pheasant
Hud-hud	hoopoe
Il	kite
Jel butara	Himalayan pied kingfisher
Jumiz	imperial eagle
Kabk	chakor partridge
Kaindal	hill partridge
Kalesur	kalij pheasant
Kalij	kalij pheasant
Kali-pholia	white-capped redstart
Kaljit	Himalayan whistling-thrush

Kangskiri	spotted dove
Kastura	Himalayan whistling-thrush, grey-winged ouzel
Kasturi	grey-winged ouzel
Koak	koklas pheasant
Koin	Indian turtle-dove
Kokia-kak	Himalayan tree-pie
Kokla	kokla green-pigeon, koklas pheasant
Koklas	koklas pheasant
Kolsa	kalij pheasant
Krishen-patti	blue-headed rock-thrush
Kuil	koel
Kukera	kalij pheasant
Kukku	cuckoo
Kukrola	koklas pheasant
Kupak	common hawk-cuckoo
Kupwah	cuckoo
Kyphulpakka	Indian cuckoo
Kyphulpakki	Indian cuckoo
Machi bagh	Himalayan pied kingfisher
Madana suga	slaty-headed paroquet
Maina	myna
Miouli	great Himalayan barbet
Mohrhaita	changeable hawk-eagle
Moraugi	Bonelli's eagle
Neoul	great Himalayan barbet

Nilkant	blue magpie
Niltau	rufous-bellied niltava
Okhab	lammergeyer
Pahari maina	jungle myna
Pahari tuiya	slaty-headed paroquet
Painju	white-cheeked bulbul
Panduk	dove
Patariya masaicha	grey-winged ouzel
Perki	dove
Peunra	hill partridge
Phupu	cuckoo
Pilak	oriole
Plas	koklas pheasant
Pokras	koklas pheasant
Popiya	common hawk-cuckoo
Puli	spotted wing
Ram chakru	hill partridge
Roli	hill partridge
Sadal	changeable hawk-eagle
Safed gidh	scavenger vulture
Sahili	scarlet minivet
Sahim	ashy drongo
Sakdudu	hoopoe
Satangal	imperial eagle
Shah bulbul	paradise flycatcher

Sibia	sibia
Sim kukra	woodcock
Sim tital	woodcock
Takpo	Indian cuckoo
Toitru fakhta	little brown dove
Traiho	great Himalayan barbet
Tuktola	Western-Himalayan scaly-bellied green woodpecker
Turkan	Western-Himalayan pied woodpecker
Tusal	bar-tailed cuckoo-dove
Tutitar	woodcock
Ulak	corby
Zakki	brown flycatcher
Zird phutki	grey-headed flycatcher

II. VERNACULAR NAMES OF NILGIRI BIRDS

Adavikodi	grey jungle-fowl
Adavi nalla gedda	black eagle
Adiki lam kuravi	sparrow
Boli kadi	white-breasted water-hen
Boli kodi	moorhen
Buchi gadu	white-breasted kingfisher
Buruta pitta	Indian skylark
Chandul	crested lark
Chilluka	paroquet

Chinna ulanka	wood sandpiper
Chinna wallur	shikra
Chitlu jitta	Nilgiri flower-pecker
Chitti bella guwa	little brown dove
Dasari pitta	scimitar-babbler, fantail flycatcher
Garud alawa	Brahmany kite
Garuda mantaru	Brahmany kite
Gola kokila	pied crested cuckoo
Goranka	common myna
Gudi konga	paddy bird
Guli gadu	white-backed vulture
Gurapa madi jitta	Indian pipit
Jali dega	shikra
Jambri kodi	moorhen
Jitta kodi	red spear-fowl
Jutu pitta	crested lark
Kadai	painted bush quail
Kakka	black crow
Kakki	black crow
Kakkara jinuwayi	spotted munia
Kalli kaka	crow-pheasant
Kalu prandu	kite
Kaltu koli	grey jungle-fowl
Killi	paroquet
Kokku	paddy bird

Konda lati	red-vented bulbul
Kumpa nalanchi	pied bush-chat
Kundeli salawa	Bonelli's eagle
Kutti pitta	hawk-cuckoo
Lak muka	white-breasted kingfisher
Likku jitta	tailor-bird
Machayarya	fantail flycatcher
Malla gedda	kite
Manam badi	Indian skylark
Manati	fantail flycatcher
Manju tiridi	scavenger vulture
Meta kali	Indian pipit
Namala pitta	scimitar-babbler
Nella borawa	Pondicherry vulture
Niala pichiki	Indian skylark
Nila buchi gadu	common kingfisher
Papa	scavenger vulture
Papa parundu	scavenger vulture
Paria prandu	kite
Pedda sida	jungle babbler
Pigli pitta	red-vented bulbul
Pit pitta	ashy wren-warbler
Pittri gedda	scavenger vulture
Poda bella guwa	spotted dove
Puli pora	spotted dove

Rajali	Bonelli's eagle
Sarrava koli	red spur-fowl
Sowata guwa	little brown dove
Tangada goranka	pied crested cuckoo
Tella borawa	scavenger vulture
Than kudi	sunbird
Tinna kuruvi	spotted munia
Tondala doshi gadu	kestrel
Tondala muchi gedda	kestrel
Tonka pigli pitta	paradise flycatcher
Torra jinuwayi	red munia
Touta pora	little brown dove
Turaka pigli pitta	hill or red-whiskered bulbul
Uri pichiki	sparrow
Vichuli	white-breasted kingfisher
Wal konda lati	paradise flycatcher
Yerra belinchi	rufous-backed shrike
Yerra kodi	red spur-fowl